Inside Mathforum.org

The internet has dramatically transformed social space and time for many people in many different contexts. This dramatic warping of the social fabric has happened slowly over time as digital technologies have evolved and internet speeds have increased. While we are all aware of these changes, the impact is often little understood. There are few monographs about social groups made possible by the internet and even fewer about educational communities made possible through digital technologies. *Inside Mathforum.org* details the ways that digital media are used to enhance the practices in which teachers and students engage. The book also shows how different kinds of mathematical conversations and interactions become possible through the digital media. Unlike many other educational uses of digital media, the Math Forum's community has provided online resources, sustained support for teachers and students, and leads the way in showing the power of digital media for education.

WESLEY SHUMAR is an anthropologist at Drexel University. His research focuses on digital media, math education, and virtual educational communities. Since 1997, he has worked as an ethnographer at the Math Forum. Currently he is Co-Principal Investigator on EnCoMPASS, a four-year National Science Foundation (NSF) project designed to build an online community of math teachers through formative assessment and a focus on student problem solving. He was the Principal Investigator on two other NSF projects, the Online Mentoring Project and the Math Forum's Virtual Fieldwork Sequence. He is co-editor with K. Ann Renninger of *Building Virtual Communities: Learning and Change in Cyberspace* (Cambridge University Press, 2002).

Inside Mathforum.org

Analysis of an Internet-Based Education Community

WESLEY SHUMAR
Drexel University, Philadelphia

CAMBRIDGE
UNIVERSITY PRESS

CAMBRIDGE
UNIVERSITY PRESS

University Printing House, Cambridge CB2 8BS, United Kingdom

One Liberty Plaza, 20th Floor, New York, NY 10006, USA

477 Williamstown Road, Port Melbourne, VIC 3207, Australia

314-321, 3rd Floor, Plot 3, Splendor Forum, Jasola District Centre, New Delhi - 110025, India

79 Anson Road, #06-04/06, Singapore 079906

Cambridge University Press is part of the University of Cambridge.

It furthers the University's mission by disseminating knowledge in the pursuit of education, learning and research at the highest international levels of excellence.

www.cambridge.org
Information on this title: www.cambridge.org/9781316503676
DOI: 10.1017/9781316481752

First published 2017
First paperback edition 2019

A catalogue record for this publication is available from the British Library

Library of Congress Cataloging in Publication data
Names: Shumar, Wesley.
Title: Inside Mathforum.org : analysis of an Internet-based education community /
Wesley Shumar, Drexel University, Philadelphia.
Other titles: Inside Math Forum
Description: New York, NY : Cambridge University Press, [2017] | Includes bibliographical
references and index.
Identifiers: LCCN 2017009164 | ISBN 9781107138858 (Hardback : alk. paper)
Subjects: LCSH: Mathematics–Computer-assisted instruction. | Math Forum. |
Computer-assisted instruction.
Classification: LCC QA11.5 .S58 2017 | DDC 025.06/51–dc23 LC record available at
https://lccn.loc.gov/2017009164

ISBN 978-1-107-13885-8 Hardback
ISBN 978-1-316-50367-6 Paperback

Contents

Figures

Acknowledgments

This project has developed over many years, and I now have had a very long-term relationship with the Math Forum. Few ethnographers get to work so closely with their "informants" over such a long period of time, such that the nature of the research relationship is blurred. I am both a collaborator and an ethnographer with the Math Forum. I owe a tremendous debt of gratitude to the Math Forum staff members, current members as well as those who have moved on to other jobs, for their support for so many years. I feel that I have learned so much about learning, technology, and mathematics from this talented group of people. I have also had wonderful collaborations with other researchers who have collaborated with the Math Forum along the way. There are too many of them to mention by name, but I am grateful for all the ideas we have developed and shared together. I would like to thank my close research collaborators at Drexel University, Valerie Klein and Jason Silverman. My weekly meetings with them and our graduate students over the last several years have been the highpoint of my days at Drexel. Their insights about the Math Forum have been invaluable to my understanding. A special thanks to my friend and colleague Jonathan Church, who carefully read the manuscript. His anthropological perspective has really helped me over the years to think about what I have been seeing in the work with the Math Forum. I would also like to thank Carol Brandt. Her thoughtful engagement with the ideas I have been working from both an ethnographic and educational perspective have helped to guide the book. She contributed significant editorial support of the final manuscript as well as moral support throughout the whole project. This book would not be possible without her.

I thank and acknowledge the following for permission to reprint: Bloomsbury Publishing for use of excerpt from Daniel Miller and Don Slater, *The Internet: An Ethnographic Approach*, Berg Publishers, 2000, used by permission of Bloomsbury Publishing Plc.; Polity Press and Stanford University Press for use of excerpt from *The Logic of Practice* by Pierre Bourdieu, translated by Richard Nice Copyright © 1990, Polity Press. All rights reserved. Used by permission of the publisher, Stanford University Press, sup.org.; and Harvard University Press for use of excerpt from *Identity and Agency in Cultural Worlds* by Dorothy Holland, William Lachicotte, Jr., Debra Skinner, Carole Cain, Cambridge, Mass.: Harvard University Press, Copyright © 1998 by the President and Fellows of Harvard College.

I would also like to thank the Math Forum and the National Council of Teachers of Mathematics and acknowledge their permission to use the following materials: the description and associated image of Problem of the Week Problem #16027, Well, Well, Well; the description of the Problem of the Week Scoring Rubric; the description and associated image of Problem of the Week Problem #4036, Math Club Mystery; a Solution to Math Club Mystery Problem; Teacher Mentoring a Student Doing the Math Club Mystery Problem; example of an Ask Dr. Math Reply to a Math Question; Approver's Conversation with Mentor for the Math Club Mystery Problem; Seven Congruent Rectangles Problem in the Online Mentoring Guide; runners graph created using NCTM 1998 online applet; and text from Math Forum Mathtech workshop page.

1   Introduction

This book is an ethnography of the Math Forum (mathforum.org), an online math education resource center that is part of the National Council of the Teachers of Mathematics (NCTM). I began doing research with the Math Forum in 1997 when it was at Swarthmore College. The Math Forum began as a project of a faculty member at Swarthmore and an undergraduate math major. Originally the project was called the "geometry forum" and it was a discussion list on the Usenet. The geometry forum began before the web and so all interactions were text-based. These early pioneers thought that the internet could be leveraged to allow individuals who were interested in problem solving and talking about math to have more conversations with each other and to share resources. As part of this work, the team worked with the developer of geometer's sketchpad. This was a software tool that allowed people to visualize geometry in this text-based environment. Sketchpad is still an active product today sold through McGraw-Hill Education.

In 1996, the group received a one-year "proof of concept" grant to create the Math Forum from the National Science Foundation (NSF). Following that funding, the Math Forum then received its first three-year NSF grant to build out the internet's first online math education resource center and community. Of course by this point, the web was born, and the visualization of that virtual space was greatly enhanced. The Math Forum then came into being in the early heady days of thinking about "online communities" and the utopian excitement about web 1.0, as it were. From these early days at Swarthmore College, the Math Forum then has gone on a long adventure of development moving from the College to WebCT to Drexel University and now to NCTM. In the process it has built one of the most long-lasting education communities (or cluster of communities) and a robust set of resources on the web for math education. The Math

Forum began as an online education community during web 1.0 and has persisted and evolved to be a node of interaction among math educators in web 2.0. As such it is one of the longest lasting online educational communities. Its members have also been pioneers of the kinds of inter-actions and communications that the internet makes possible and how these new forms of interaction can be leveraged to help teachers and students of mathematics improve the work that they do and the know-ledge they produce.

I originally began working with the Math Forum as an ethnographic evaluator and was on their original three-year NSF grant as an evaluator. I moved to many different roles with the Math Forum. I eventually became principle investigator on several projects, as well as continuing to work as a researcher and evaluator on others. I used to jokingly suggest that the Math Forum had become my main field site and that math was the language of my tribe, but of course this was an ironic illusion, as the metaphor had many difficulties. Many anthropologists have discussed the dilemmas of ethnography in the contemporary world, and ethnography with an online math education community has its own particular compli-cations. Gupta and Ferguson (1997) suggested, at the same moment that I began working with the Math Forum, that the "field" was a concept used by ethnographers and that it needed to be deconstructed. Fieldwork, according to them, implies two spaces, the field and home, and, more important, two kinds of writing, fieldnotes and finished ethnographies, each done in their respective places. Likewise George Marcus has done a significant amount of work recently focusing on ways that globalization has impacted the practice of ethnography, as groups are no longer spatially contiguous and what binds them is very different from the traditional notions for community (Faubion & Marcus, 2009; Marcus, 2010; Marcus, 2012). These are all issues I faced with the Math Forum, where it was difficult to tell what were fieldnotes and what were writings for other purposes. All of these ideas I'll explore in more detail in Chapter 2 of this text.

One of the first activities I engaged in at the Math Forum with the staff members was an activity we called "mapping the Forum." I will talk about mapping the Forum in more detail in Chapter 2, but for here, it's important to point out that this activity was about making visible the more invisible elements of a social structure that was organized through digital media technologies. I quickly came to trust that the Math Forum was a dynamic community of teachers, students, researchers, hobbyists, and parents. And this community had a number of dynamic projects going on, interesting collaborations, deep discussions, and "capital" in the form of math and teaching resources. But, for the most part, these things were not observable, and even when they were observable it was on a computer screen, through which one had to know how to browse or search. An important task for me as an anthropologist, but also valuable to the staff themselves, was to "objectify" these practices, to make visible the barely visible so that people could reflect on them more productively.

While originally a project, over time the Math Forum (mathforum.org) began to think of itself as a virtual math education resource center. It was, until recently, housed at Drexel University where I currently work. And it was at Drexel for a long time. We each came to Drexel at different moments, I in the academic year of 1996, and the Math Forum in 2001. The Math Forum left Swarthmore College and was bought by WebCT in April 2000 during the tail end of the dot-com boom. It was then "spun off" by WebCT during the dot-com bust. Drexel University picked up the Math Forum shortly before they would have been force to close down operations – that was in June 2001. This co-location was fortunate for me, as it allowed me to have a much deeper and more intimate relationship with the staff at the Math Forum than I might otherwise have had. During their time at Drexel, the staff of the Math Forum occupied one room at the university. It is a largish space with staff people who run the services, software developers who produce and maintain the site, and a back room with a bank of servers that house the virtual world of the Math Forum. The physical location of the Math Forum and the onsite staff are an important part of what the Math Forum is. A core of people met in this office every day. But around this core group revolved a set of telecommuting staff and a large set of virtual arrangements that could really be thought to constitute a nested set of communities or subcommunities. I will discuss these organizational elements more in Chapters 3 and 4.

There are several goals I have in this monograph. On the one hand, I want to situate the Math Forum within the broader structural context of changes going on in the US economy, and especially the internet economy. The optimism about internet organization when the Math Forum began influenced the plans for how the Math Forum would develop. And the dot-com bust had a big impact on the Math Forum and where it was able to go. Like many anthropologists, I want to use this macrostructural contextualization in order to frame the activities that have gone on within the Math Forum. My second goal is to look at the contributions the Math Forum has made to math education. Its focus on math as a practice that all can engage in and that all can talk about is an extremely important model. A third goal is to look at the contribution the Math Forum has made to thinking about how to use internet and digital technology for learning. The Math Forum has kept its sights on interaction and problem solving, and this has helped it to think about the things technology can contribute to people who are learning math. While I will look at some of the things students do at the Math Forum, my main focus in the volume will be on teachers. Teachers are the people who spend the longest periods of time with the Math Forum, and they form most of what I would call the online community that is the Math Forum.

This volume exists within a small set of similar case studies. There are very few ethnographies of online educational communities. In education there are a number of edited volumes that have chapters discussing various aspects of digital media and online educational communities or groups (Barab et al., 2004; Ching & Foley, 2012; Falk & Drayton, 2009; Renninger & Shumar, 2002). Each of these edited volumes has some important case studies on different online educational groups. Further, they also explore some important methodological issues as well as important conceptual issues. Ching and Foley's (2012) volume, which is the most recent, has different chapters that engage with the important idea of identity and the ways identities are constructed and transformed online.

One of the few existing case studies is Slotta and Linn's (2009) book on the WISE project in science education. WISE is a collection of curriculum projects that individuals in different locations might use and so the book is really about the sharing of resources and the kind of distributed community that forms around these resources. As such, the WISE shares a good bit with the Math Forum, which is also a resource site for sharing assets. The WISE book is more a report on the use of the WISE curriculum and the ways people have connected with it. It never had the intention of being an ethnography. This volume, I hope, will fill a significant gap in the literature.

My research is influenced by learning sciences researchers and math education researchers who have taken a more social learning theory approach (Boaler, 2000; Cobb et al., 2000; Sfard, 2008). Some of these researchers have discovered ethnography and have been using ethnographic theory and techniques to enhance their work (Barab et al., 2004; Bell, 2012; Gee, 2007). This work has been very important in that ethnography has informed curricular design as well as giving researchers in education a good sense of the importance of looking at learning practices within larger social context. This research using ethnography probably goes back to the work of bringing Vygotsky to psychology and the general work in cultural psychology (Cole, 1998; Rogoff, 2005). In a related way practice theory in education and the notion of community of practice has been important to thinking about online education groups (Bourdieu, 1990; Lave & Wenger, 1991; Wenger, 1998). Similarly, Gee's (2007) research on sociolinguistics and learning has been critical to the learning sciences and education research in general. Gee's theoretical ideas, like affinity spaces, are critical to the development of an ethnographic perspective in this area. Further I would argue that *What Video Games Have to Teach Us about Learning* is primarily an ethnographic text, although Gee frames it more as a kind of semiotic analysis.

Likewise in anthropology there have not been a large number of ethnographies of online communities. Boellstorff's (2008) *Coming of Age in Second Life* is probably the main text that one could point to as an ethnography that stays within a bounded virtual world and makes an interesting argument for doing so. He also briefly discusses some of the "ethnographically informed" case studies in this area, but again there are not a large number of them and many of them were written in the 1990s. Boellstorff also takes on a number of the interesting questions about virtual/actual/real, the questions of place in ethnographic research, the fluidity of otherness, and so on. It is an important work in this area. But in some ways virtual worlds like Second Life or online gaming worlds makes this a simpler matter because one can talk about the experience of natives and how the ethnographer shares that by remaining focused on the virtual space and the life in that space.

Miller and Slater (2000) attack the issue of virtual community in a different way by suggesting that a case study of the internet could in fact be the case study of any group of people and the ways they are able to reimagine themselves with new communication technology. In some ways Miller and Slater's book is about how Trinidad is a different place and Trinidadians are different people thanks to the internet. The internet has allowed for a warping of the social fabric that was previously not possible. Not only can Trinidadians on Trinidad think about their relationships with each other differently, but Trinidadians who have emigrated to other countries can maintain a homeland identity in a way that was previously not possible. Trinidad is not only a nation, but it can become a space of affinity (Gee, 2005). This study of the Math Forum shares some important perspectives with Miller and Slater's work. The Math Forum has woven together a complex social space within which bonds of affinity can take place, and this has created multiple and overlapping groups. While the Math Forum is not a country and its people do not share a nation, there are similar ways to the ways Miller and Slater discuss that the Math Forum has helped to warp the social fabric. If Anderson (1991) has taught us that communication technologies are often at the core of a group's imagination, new communication technologies have allowed us to understand spaces and groupings and even temporality in heretofore unimaginable ways.

New Locations

As soon as groups of people began to congregate through new technologies they found themselves in a new land, cyberspace. And immediately spatial metaphors began to abound. The internet created the possibility for all kinds of communicative interactions, but it also did more: it created persistence. This is something that one can see developing even before the internet. Perhaps for consumers, some of the first social cyberspaces were the spaces created on one's phone answering machine. But this space indeed was limited. And as social cyberspace became more common, people began to talk about online community and virtual community. And there has been a long discussion about community, online communities, hybrid communities, and the transformation of all communities. I've found myself caught up in this discussion at a number of historical junctures (Renninger & Shumar, 2004; Shumar & Renninger, 2002). The debates in anthropology about spaces, locations, virtuality, and the digital and the hybrid have yielded some interesting review articles (Coleman 2010; Shumar & Madison, 2013; Wilson & Peterson, 2002). What is critical in these discussions is not specifically whether we are talking about a community or something more like new forms of affinity but rather that we have a way to theorize the social space and how social life and its practices are organized temporally and spatially, whether these spaces are completely digital, virtual, or actual (Boellstorff, 2008).

From the perspective of these debates about the virtual and the digital the Math Forum is a collection of resources (lesson plans, math problems, FAQs, etc.), a few services (Ask Dr. Math, the Problem of the Week), and a set of discussion lists. But because of the nature of internet communication technologies and the ways people use them, the Math Forum could be conceptualized as a community or a community of communities. And as such it involves people who interact in distributed ways as well as face to face. There are regular participants on the Math Forum site who participate across a number of different lists and services. But there are also regular participants who are just part of one area (e.g., Teacher to Teacher (T2T), a discussion group around issues of pedagogy). People who see themselves as members of this discussion group might have little involvement in other parts of the Math Forum site. The core of what makes the Math Forum a live sociality is the persistence of resources and the persistence of the traces of earlier conversations. These archives make future participation possible and are what people are looking for in this organization.

Participation at the Math Forum is interesting precisely because it demonstrates the transformative potential of the internet for individuals and communities. Coleman (2010) suggests that anthropologists are skeptical about the life-transforming nature of the internet. That certainly makes sense. We know that the internet has changed things, but like other communication technologies, it has been integrated into the practices that already went before it. What is hard to value are the tectonic shifts that the internet has brought about. We tend not to see them because they unfold through time, and the past is always a foreign country (Lowenthal, 1999). There are several unique features of the Math Forum. First, it is one of the oldest, best-known, and most active online educational communities. It has been challenged, but not terminated, by changes in the forms of support and institutional locations it has experienced. Although at times it looked like things were going to end for the Math Forum, it always managed to bounce back. It has continued to inspire teachers and students to focus on the everydayness of math, the pleasure of problem solving, and the importance of talking and thinking about math. It is truly a remarkable organization.

Themes

In this section, I will talk about some of the key themes that will come up in the book and form important parts of future chapters. These themes are interlocking, but I lay them out with separate headings for analytical purposes. This will allow the reader to think about what is often implicit in a chapter in a more concrete way.

Math Forum Culture

In anthropology in recent years the concept of culture has been criticized. A number of anthropologists have suggested that we even discard the notion of culture. But I would maintain that the notion of culture is useful for thinking about the Math Forum (Renninger & Shumar, 2004). Following Holland et al. (1998), I would suggest that the Math Forum culture is a process and one that is still unfolding. It is caught up in the practices of the Math Forum and the social imagination of the members of the Math Forum community. My understanding of culture is very similar to the notion of "figured world" in that it is a process, is made up of strands from other processes, is structured and is caught up in the material relations within it and surrounding it, and, importantly, involves the creative actions of the individuals involved (Holland et al., 1998: 60).

For simplicity's sake, we can think of the Math Forum culture as having three major influences: the small liberal arts college where it began, the utopian culture of the early internet, and the dynamic personalities of the founding members. These influences, of course, produced their own history, which built the culture and is constantly being reworked at each new juncture.

Swarthmore College is a thoughtful and well-resourced environment. Students are talented and they tend to share a progressive orientation toward education and the world. They also have a habitus that makes them resilient problem solvers. The early Math Forum staff was made up of a number of former graduates of the college or similar institutions. While the members of the staff did not come from the most elite family backgrounds, they shared a social class and educational privilege that shaped the early culture. They were and are thoughtful, engaged in the world, and problem-solving oriented, and they cared about improving educational opportunities for a wide range of students and teachers.

The early culture of the internet nicely dovetailed with the progressive orientation of the College. Markoff (2006) discusses some of the ways that the personal computer industry grew out of the 1960s counterculture in the San Francisco Bay area. That culture, he suggested, carried with it progressive ideas about the freedom of knowledge and information and the ways that could make a democratic and more egalitarian society. The culture of the internet was very much connected to this utopian '60s counterculture. The internet would not only allow for free and unfettered communication between individuals, it would also bring digital goods that could be distributed freely and begin to demonstrate that a society based on plentitude, not scarcity, was possible. Poster (2001) continued to take up these ideas in his book *What's the Matter with the Internet?*, suggesting that this utopian potential of the digital was very much at odds with our economy based on scarcity. In certain arenas we continue to see this tension around the possibility of plentitude. In a real way the Math Forum took advantage of the utopian notion of open and free discussion and the sharing of educational resources. Because they were dealing with math, and not something like popular music, the pushback on their views was different. They were pressured to commodify their resources so that they could be self-sustaining, and this had a limiting effect on the vision. But we will discuss that more in Chapters 3 and 4.

The third element of the culture is the founding members of the culture, as well as the individuals who join the Math Forum later. The Math Forum is made up of a group of people who as participants of the small liberal arts college culture and/or the early utopian internet culture shared the optimism of how the internet can benefit education. But more specifically, the members of this culture, each in their own way, had an interest in people and an ability to take people for who they are and for where they are located. They were all genuinely interested in the well-being of others, and they had and still have an ability to see past a person's social status when working with them. The Math Forum staff took advantage of the internet's capacity to suppress a person's status. When talking with teachers and students, the Math Forum staff are interested in who that person is and in having a conversation with them, regardless of where that person came from. One of the early mentors in the Problem of the Week (PoW) system was a teenage student who just happened to be very good at math. And the student not only worked as a mentor for math fundamentals, this student was a mentor at every level of mathematics. The Math Forum staff members did not see this person as a category (teenage student) but, rather, as a person who was not only very good at math but very good at talking about math. They cultivated that person because they are interested in people and good conversations. They were, of course, interested in helping that student move even further in their own thinking.

Math Forum Dialectic

I came to see the way the Math Forum approached mathematics as a dialectic. The focus at the Math Forum starts in practice and is very much about problem solving and doing math. Math Forum staff think constantly about doing math because they see math as rooted in the everyday. And while some people may need to do more math and more complex math than others, math is part of everyone's life. In order to get better at doing math not only does one need to engage in the doing, but one needs to talk with others about the math that is being done. Finally, doing and talking leads to ways of thinking, which then lead back to practices. This process is ongoing. It may involve short-term reifications, like writing things down and taking notes, but there is never really a final reification – the answer. Answers lead to new thought and new questions.

When I began to see how central communication was to the Math Forum and the process of doing math, it made me think of Peirce's statement that all thought is dialogic (Peirce, 1931; Shumar & Madison, 2013). I have been using Peirce's point for a long time to make the link between thought and communication. Later I discovered the work of Anna Sfard (2008), who addresses this connection between thought and communication brilliantly. The Math Forum's original interest in the internet was that it created more and different opportunities for talking about math. And because it allowed these opportunities, one could get to know the other person better and have a much better sense of how they thought about mathematics. This sense of how others think is invaluable. Understanding how learners think gives one ideas for how to mentor that person and help them move their thinking about math forward (Pea, 2004). But, of course, the mentor might learn new things him- or herself in that process and become better at mathematics. The Math Forum has always seen math education in this dialogic way. It is not a matter of teaching kids things; rather, it is a matter of having a conversation where each participant in the conversation might move his or her thinking. The Forum approaches students, teachers, and other colleagues in the same dialogic fashion.

The fact that these conversations can be technologically mediated allows for more and different conversations and opportunities for doing math, as well as allowing for former bits of dialogue to be incorporated into new interactions. In a very real and practical way the Math Forum has always seen "utterances" in a Bakhtinian way: discussions are built around new speech acts, but they might potentially incorporate the text of former utterances as well. Interestingly, these technologically mediated conversations opened up new technologically mediated spaces. The Director of the Math Forum always said to me that community was not something the Math Forum was seeking. Rather, it was an effect of the effort to improve opportunities to do math and to have conversations around mathematics.

If we think of the notion of reification – turning activities into things – and the central role that reifications play in human thought (Sfard, 2008), the Math Forum produces different kinds of reifications and interacts with them differently from what we might expect given more traditional ideas about math education. If the traditional math classroom was once focused on correct answers and on the procedures and mechanics of doing mathematics, the Math Forum is not particularly interested in reifying those parts of the process. It is interested in learning to think mathematically, and so good conversation and interesting problem-solving activities are the important parts of that process for it. Certainly, good procedures, clear documentation of work, and correct answers are all part of these interactions. But they are only important in that they are evidence of a thinking process. What gets reified for the Math Forum are good ideas, interesting strategies, interesting incorrect approaches, good conversations. They are reified so that they might be used again in other interactions to move the conversations and thinking forward in future problem-solving activities; that is, they are reified in the service of future process.

Double Reflection

An important part of the dialectical process is something that I have called "double reflection." We could think of this as a form of consciousness or maybe a subject position as well. Double reflection is a process that the Math Forum engages in; it's not what its members would call it, but it's a name I have given to a Math Forum process. It's a core part of what was eventually named "Noticing & Wondering." And these ideas and processes are similar to others that we see in the math education literature as well as in social learning and constructivist theory more generally. Double reflection to my mind is a collaborative knowledge-building activity. It begins with teachers, mentors, or Math Forum staff engaging in the process of problem solving and the discussions they have with others about that process. A core part of the conversation involves reflection on the problem-solving process itself. It's a meta activity, as one is not only trying to solve a problem but is reflecting on how one engages in that problem-solving activity – what did I think of first, what did I try to do, what did I get wrong, what did I get right, what else could I have done, and so on.

The second stage of this reflection is when one talks to someone else who is solving the same problem or a similar problem. It begins with an effort to genuinely engage with another and both understanding what they are saying and attempting to imagine the thought process that went into what the person did and what they said about what they did. One draws on one's own reflection in this problem work to engage in this second reflection. And questions and discussion are about both trying to get a better grasp of what is going on in the other's head – not just of the informant, student, mentee, or colleague – and moving the thinking forward – where the mentor might also develop new ideas and advance his or her thinking. It reminds me a bit of the anthropologist's effort to take the other seriously. The Math Forum staff use this process of double reflection with each

other, other teachers, students – really, anyone. I have many times experienced this process and had this amazing guidance of my own mathematical
thinking. I made many interesting realizations where there was something
I thought I knew, but I knew only a procedure; I did not really understand
the math.

Organization of the Book

Chapter 2 focuses on my understanding of an ethnographic perspective.
I was trained in a traditional anthropology department and so many of my
touchstones about the problems and issues of ethnography come from
researchers who have not done any research in education. And while
I would now say my closest colleagues are educational ethnographers from
a variety of fields, I tend to have a core set of ethnographic concerns that
come from the field of cultural anthropology in the American context.
As more of the world is drawn into collaborative and global social relations,
ethnographers have responded to these challenges by asking questions
about the traditional division between "home and field" and how one works
as an ethnographer collaboratively with other researchers and with one's
informants. Further these hybrid spaces are having an impact on our sense
of the communities to which we belong and the identities that we inhabit
within these social groups. The chapter will discuss the importance of the
question of social space and engage with some of the education literature
on social space. The chapter deals with the important questions about
the role of social space in the construction of groups, the manipulations
of modes of communication, and the implications for belonging to a social
group and identity.

As a technology organization, the Math Forum has transformed its own
social space and social world, bringing together physical spaces with virtual
spaces, creating a complex social fabric. In this way, it is like a number of
contemporary organizations where new information technologies not only
have allowed for the complication of physical space (working on site,
telecommuting, etc.) but have also encouraged the transformation of social
space by flattening hierarchies and creating a creative organization where all
voices have access to the productive creation of new ideas. It has been an
ideal organization for thinking about some of the contemporary problems
of ethnography.

Chapter 3 looks at the history of the Math Forum and how it grew from
a small group of faculty and students inhabiting one room at Swarthmore
College to a much more complex organization now part of the NCTM.

The Math Forum began as an NSF project in the early days of the Usenet. At that point it was the geometry forum and was the project of a math professor and an undergraduate student in mathematics. As the geometry forum moved to pilot the idea of a "math forum" it became a group with a small staff that occupied a single room in a small liberal arts college with a Quaker tradition. The chapter uses the different physical locations that the Math Forum has inhabited, and the ways those contexts have affected them, to organize the first part of the history. The history of the Math Forum is then told from the perspective of the major projects of which it has been either a part of or the originator.

Through the lens of locations and projects the chapter will talk about the ways in which the Math Forum used hybrid social spaces (face-to-face workshops and online meetings and groups) to build a core community that could be called the Math Forum community. This history not only clearly defines the growth and development of the Math Forum but marks several critical moments. It shows how learning organizations are truly different in the internet age and the many ways this organization could be a model for the future. The Math Forum has been a unique organization in the ways it has been able to take advantage of the different physical contexts in which its members found themselves. The Forum also has had a series of projects that have unfolded over years that has allowed it to develop its vision of math as a regular part of life and how reflecting on one's thinking can truly change how people understand math.

Chapter 4 builds conceptually on Chapter 3. The original Math Forum project had two idealistic ideas built into it. First, it was to be self-sustaining. In that early moment of internet culture, people believed that the internet itself could somehow sustain organizations since communication was inexpensive and digital resources could be reused. In our modern world of massive server farms, this belief could be seen as quite naïve. Second, it was to be a model of other online educational communities. It was thought that the creative ways the Math Forum had leveraged new technologies could be taught to other groups.

If much of the Math Forum practice was an organic and unalienated outgrowth of the idealism of liberal arts education and the utopian culture of the Internet, then that unalienated labor came into direct confrontation with the ideology of the commodification of the digital educational economy. This chapter looks in greater detail at the ironies of a neoliberal view of education in the digital age. The chapter has implications for the current practices of online education in universities and new products such as massive open online courses (MOOCs). The chapter looks at tension between the tremendous resilience of the Math Forum and what that implies for the potential of other internet educational organizations and how the growing commodification of the internet might have limited what the Math Forum could have become.

Chapter 5 also develops ideas from Chapter 3. The chapter focuses on the core of the Math Forum culture and its focus on doing problems, sharing in mathematical conversations, and the building of mathematical ideas through this process. The chapter discusses a tradition at the Math Forum that existed for several years called Math Monday. The Math Forum staff met on Mondays, both face to face and also with remote participants, and did math while discussing the work of doing math and generating math problems. Central to the Math Monday process was a wiki where all the work, notes, and activities were kept. Math Monday was one of the ways the Math Forum forged itself into a "community of practice" and how that community of practice produced mathematical objects to further discussion and thinking. The chapter develops the Math Forum dialectic, a key theoretical practice that involves three foci: doing problems, talking about the mathematical work, and then thinking about mathematics. This theoretical practice informs not only the staff's own work but how they work with students and how they structure professional development to support teachers. The chapter discusses how the Math Forum's three-part practice of doing problems, talking about mathematical work, and trying to advance one's thinking about mathematics is very much related to contemporary ideas about learning in the learning sciences and related fields. The chapter concludes with how theory is related to practice and how the Math Forum practice attempts to change the culture of fear and avoidance with mathematics that one finds in so many educational institutions.

Two key services the Math Forum developed in the early years of its history were Ask Dr. Math, a question and answer service, and the Problems of the Week (PoWs), which were a series of nonroutine challenge problems available to students weekly. Both services focused on the use of technology to mentor students of mathematics. Chapter 6 discusses the development of both services but focuses on the development of the PoWs and how important they have been for working with students and faculty at the Math Forum. The PoWs began with the Geometry Problem of the Week (Geo PoW) but expanded to a suite of different problems. The development of these problems is the central activity of Math Monday. Critical to the Math Forum was the development of a way of scoring the problems and providing feedback for problem solvers.

From this work on developing an interactive problem service the Math Forum developed both a set of online courses for in-service teachers and a mentoring service for pre-service teachers. The chapter discusses several projects in which the Math Forum is engaged. The Leadership Development for Technology Integration (LDTI) project was one project where the Math Forum created a set of online math courses for in-service teachers using the PoW service at its core. The project then moved to develop teacher leaders who would contribute to the development of new online courses. The chapter also discusses two pre-service teacher projects, the Online Mentoring Project (OMP) and the Virtual Fieldwork Sequence (VFS). Both of these projects used the PoW to teach university students how to mentor pupils in the PoW and as a way to encourage the development of mathematical conversation and mathematical thinking among pre-service teachers. A critical piece of these projects worked to bring the culture of the Math Forum, with its focus on problem solving, to the cultures of several schools of education where faculty told us they were hopeful that the Math Forum culture could be influential. Throughout this work with mentoring students and faculty, two key ideas developed at the Math Forum. First, the serious focus on student learning and listening to student's ideas. Digital technology allows Math Forum staff to slow down the process of working with students and listen for the interesting thinking, even when students are articulating wrong ideas. These interesting ideas can be the way to improving student learning and helping them to think mathematically. Second, the Math Forum realized that digital technologies and online spaces can be used to produce teacher professional development that is more sustained, interactive, and targeted toward teachers' needs.

Chapter 7 returns to thinking about the idea of the Math Forum culture. The Math Forum, with its emphasis on developing relationships with people in order to have deeper conversations about mathematics and to do problem solving together, evolved a discursive structure called Noticing & Wondering (N&W). N&W could be thought of as a scaffold that makes it easier for individuals to engage in the process of double reflection, talking about one's own problem solving and then using that to talk about other people's problem solving. N&W has worked as a scaffold for helping teachers to engage in this reflective process. But it is also a scaffold that helps students, or any problem solvers, think about what is going on in a math problem.

The chapter moves from a general discussion of N&W and its relationship to other comments in the field to talking about a current project, EnCoMPASS, which is focused on building a community of teachers around taking students' mathematical ideas seriously using a piece of formative assessment software. Originally the project began with the notion of "rubrics" to help teachers focus on the idea of taking students' ideas seriously. But as the project developed, the teachers helped the Math Forum staff to refocus the software tool on the Math Forum's notion of N&W. The chapter documents the development of the software, the input of the teachers, and the way the software helps to move everyone to focusing on student work in more detail. The chapter concludes with some reflections the process of community development in the era of web 2.0. If the original Math Forum community was primarily located at the Math Forum itself or with other institutional partners, in the era of web 2.0, community development involves the incorporation of teachers' blogs and the Twitter universe of which teachers are part. This is a more complex two-way community development process. In this way the second part of Chapter 9 builds toward ideas in Chapter 8.

Chapter 8 begins with a discussion of persistence and the notion of space that has been so central to internet culture. Following up on ideas in Chapter 2, the ideas of space and social space and their central role in the life of communities and cultures are discussed. The chapter then goes on to talk about the digital library movement and the way that the Math Forum discovered itself as an interactive digital library in the context of the National Science Digital Library. This particular imagination of space was important for the Math Forum's vision of itself and it shaped ways that the Math Forum continued to develop resources and communities. The chapter then goes on to discuss how these imagined spaces become spaces of transformation as individuals both contribute to the space and see their own potential as a new self developing within this transformed space. Finally the chapter ends with further reflection on the notions of community, community of practice, and spaces of affinity as related concepts that each in different ways helps explain what we see at the Math Forum.

Chapter 9 picks up on the implications of Chapter 8 for individuals. Drawing on the critical insight that the internet allows not only for time/space compression but also for time/space expansion or, maybe more properly, the warping of social space, the chapter talks about how the internet allows individuals to engage in the crafting of their own social groups and the implications of that potential for people. The chapter opens with thinking about identity and the possibility for human agency more theoretically. That work is then used to think about the ways that individuals shape the social groups that they are part of and how that process has changed from the early days of the internet to the advent of social media and web 2.0. Finally the chapter concludes with the implication of this new kind of learning organization for the development and transformation of the self for both teachers and students. The uniqueness of the Math Forum as an online organization is underscored and the way that it can serve as a model to think about future educational organizations and the kinds of interactions, groups, and persons that can be part of these future organizations.

The Conclusion reviews some of the main points of the book and underscores the main dynamic tension in the book. The internet has brought significant new opportunities and potential for an education that is engaged and reflexive and gives ordinary teachers and students real opportunities to engaging in knowledge production processes. The Math Forum is a creative example of how this can happen. It is one of the very few long-lasting math education internet communities. The Math Forum, as a learning organization, also has figured out some important things about how the internet can be used to support learning, problem solving, talking about math, and teaching. But the growth of the Math Forum was very much tied up with neoliberal ideas about the ways the internet can make content easily available and Taylorize teaching techniques for the masses to consume. A commodified vision of what the internet can do for education limits the more powerful potential of the internet for education. A question one could ask is: what could the Math Forum have become at a different moment in time? And it is still possible that the Math Forum might reach a new potential in the coming decade.

Even given these tensions of the internet economy, the Math Forum has proved to be a tremendously resilient organization. Now that it is part of the NCTM it is possible that its impact on math education could be even greater than it has been so far. The Math Forum's focus on the every-dayness of mathematics, the importance of the practice of doing math and talking about math – in general its careful use of reifications and abstractions and the potential of the internet to enhance human potential – is an important model for education in the twenty-first century. The model of double reflection and the potential of digital media to enhance the opportunities for reflection and the power of reflection are again significant models for all educational practice. While many of us are still caught up in limited models of online courses and online teaching and learning, the Math Forum has been pointing the way toward a much more productive use of these tools for education.

2 Ethnography with a Leading Internet-Based Educational Center

Introduction

In 1996 I was working at a small elite liberal arts college in the Philadelphia area. I was finishing turning my dissertation on the commodification of universities into the book *College for Sale*. I was looking for another project and was also looking for something where I could report on a positive contribution to human well-being. The critique of university commodification was an interesting and rich project, but it left me, and continues to leave me, somewhat demoralized.

A colleague at the college approached me the following year and asked if I would be interested in joining her in the evaluation of a new group, the Math Forum. At that point, the Math Forum had received a one-year proof of concept grant from the National Science Foundation (NSF) to see if it could take the geometry forum, a Usenet group developed at the college, and turn it into a forum on the web for the math more generally. They were working on a larger three-year NSF grant, which they later received. And I became part of the evaluation team that was looking at the formative evaluation of this new online math education community.

I began meeting with Math Forum staff members. As mentioned in Chapter 1, I was an early adopter of the internet and personal computers, and I found them interesting and fun, but I had not formally thought about them in terms of a social space. As I began to work with the Math Forum I realized we needed to visualize this space within which its activities took place. So we began an exercise I informally referred to as "mapping the forum." The idea of mapping the forum was to get my informants to engage in a brain dump so that I could see who they worked with, what they worked on, and how these connections were established and maintained.

My colleague Marc Smith began his work on the Usenet with a similar concern, but one that was much broader than mine. Smith liked to point to the sociologist William Whyte, who used cameras and time-lapse photography to visualize patterns of crowd behavior in New York City, the same crowd behavior that at street level looked simply random. The question of making the invisible visible is one that has been important to Smith for his whole career and led to several software platforms with which he was intimately connected.[1] This is an issue that social network theorists try to deal with too. Their models allow for the visualization of networks of interaction and potential interaction. Network analysis does some important things for seeing the overall patterns of Internet use and the kinds of interactions in which individuals were engaged.

I too was interested in making invisible things visible. But I had a slightly different agenda. As much as the Math Forum was part of a larger network of connections, it was much smaller than the whole of the Usenet, and I was interested in the textured quality of the relationships that made up the Math Forum. So we began our effort to map the Forum with a "think aloud" sort of process, and that effort led to a three-part structure of teachers, partners, and projects. Each of these groups was a kind of node in the Math Forum network. This process of working with the Math Forum began with my discussions with the leadership of the Math Forum. But soon our efforts to Map the Forum became a regular part of Math Forum activities. We had regular meetings where we discussed the map and we captured different staff members' perspectives on who we were working with and what we were doing. The Math Forum used my effort to understand the social space of the Math Forum as an opportunity to deepen its own knowledge about the communities it was part of and the relationships it had with other people and groups.

Each of the elements of the Math Forum map delivered other information about the Math Forum as an organization. Teachers were the most important element in the Math Forum mapping activity. The Math Forum saw itself, from the beginning, as a resource for teachers and a place where teachers could meet, talk, and share ideas, develop their pedagogic and mathematical thinking. And in 1997 only a few K-12 teachers were actively involved in the internet, so they were a special group. Some of these teachers would eventually become key Math Forum partners. They would provide lessons and other resources for the website. They would also become part of active discussion communities and volunteer for services such as Ask Dr. Math and the Problem of the Week (PoW). They would also become the core of the Math Forum's online community.

Partners could be individuals but they were often organizations. Some-times it was hard to tell the difference between a project and a partner. But in general, partners were other organizations that were working with teachers and students of mathematics or a related STEM discipline. Some of the larger better-known partners were organizations such as TERC and the National Council of the Teachers of Mathematics (NCTM). But partners could also themselves be NSF projects that set up organizations with which the Math Forum could collaborate. Finally, partners could be individuals who were leaders in other institutions, the key players in large funded projects, or individuals who tended to have a series of related projects with which the Math Forum could collaborate. Later, not surpris-ingly, the Math Forum would collaborate on grant applications with some of these key partners.

Finally, projects were things that the Math Forum was doing. Most of the key projects in the early days were connected to services such as Ask Dr. Math and the PoWs. But also projects were discussion lists such as Teacher to Teacher (T2T), which was set up so that teachers could discuss pedagogical issues around teaching mathematics. There were projects that were separate from the Math Forum, ones at SRI International or again at TERC. These were projects for the Math Forum if the Forum was actively collaborating in the project and providing some service to the project. So the project category had both internal projects (that may have also had external collaborators) and projects that the Math Forum was collaborating on, but their organizing locus was at different institutions. All the projects had both physical and virtual dimensions.

This work on mapping the forum had an unintended consequence. As I said above, my main goal was to try and visualize the Math Forum and how we might think about it as an organization, a community, and a dynamic set of working relationships. Very practically as an ethnographic evaluator, I was trying to get my hands around the entity so that I could think about what a formative evaluation structure might look like. But the Math Forum was and is a highly self-conscious group of independent individuals, many of whom have attended elite liberal arts institutions and think critically as part of their everyday practice. The "mapping the forum" activity became a regular Math Forum activity for quite a while. The staff began to use it as a way of knowing more details about what other parts of the organization were doing. They also used it philosophically to continue to ask the question "Who are we?" And they used it to map future activity and think forward about where they could go. From the beginning of my work with the Math Forum I became part of the group. I had always been an outsider and separate from the Math Forum, yet my ideas and projects usually were taken on by the Math Forum staff, internalized as part of their practice, and then evolved, so that I became a member of the community as well.

Ethnographic Distance

It's taken me a long time to write an ethnography of the Math Forum. From my perspective, the Math Forum is a community to which I belong, a group of colleagues with whom I have worked for the past eighteen years, a website that houses a dynamic web of online groups and practices, and a business that has seen many different incarnations. I have written many things about the Math Forum and with the Math Forum and I have served as principal investigator (PI) and co-PI on many NSF projects that have grown out of the Math Forum. So it's an interesting question of why I have only now done what I was trained to do: write an ethnography of this group.

In an essay on the literary critic Erich Auerbach, Edward Said (1983) suggests that Auerbach could only write his classic tome on representation in Western literature, *Mimesis*, from Turkey. He needed the distance from the core of the culture and the libraries that housed the canon of Western literature in order to see the culture as a whole. Said had a larger point about the interconnectedness between the East and the West, that in order to frame the West, Auerbach needed a place from which to stand. But there is also in this story a more general point about critical distance. And while anthropologists have questioned the ethics of distancing oneself from one's subject, it remains true that distance (social, technological, physical) is always a kind of reflection that is difficult when one does not have the separation. This is, of course, one of the points of books such as Levi-Strauss's *Triste Tropiques* or Rabinow's *Reflections on Fieldwork in Morocco*.

The classic trope of ethnography involves two locations, separated by space and time, between which there is travel. Travel is the sacred vehicle that brings anthropologist to informant where, for the anthropologist at least, one engages in a very different time/space reality. Cut off from all of one's obligations "back home" and, in the early days, cut off from communicating with one's world, the ethnographer was free to enter into a ritual space and time that was the everyday world of the other. And as such, the ethnographer attempted to be the other, or at least simulate the other, to share their reality, their problems, the ways they got through the day, how they thought about the human condition, and so on.

For Faubion (2009: 149), this "ethic of connectivity," as he calls it, is at the core of classic ethnography and at the same time it is ethnography's problem; he sees this as both an ethical problem and an epistemological one. This is a perspective shared by many of the anthropologists of the 1990s who were engaged in the critique of anthropology (Clifford, 1988; Clifford & Marcus, 1986; Marcus & Fischer, 1986). Johannes Fabian (1983) concisely characterized this myth I briefly framed above as a denial of "coevalness" – a denial that we share a time and space with our informants, and therefore an ethical abnegation – we can use these informants for our research and our publications, but we have not responsibilities toward them.

One thing that the new world of internet spaces has done is to contribute to the destruction of this core myth of ethnography. Space and time have always been a social construction, and as Henri Lefebrve and David Harvey have taught us, space and time are usually organized around the accumulation of capital by dominant groups in the global economy, given what the material and social conditions will allow. But new technologies have allowed so much movement that the overlapping of locales has become much more common and contradictory (Appadurai, 1990; Shumar & Madison, 2013). This means that informants are moving across different groups and involved in different "collaborations," as Marcus would say, in different spaces. And ethnographers may find themselves caught up in the same set of movements or a parallel set of movements that bring new problems of how to follow groups of informants in different spaces. Nowhere is this dilemma clearer than in online spaces where defining the object of ethnography becomes quite complex (Boellstorff, 2008; Hine, 2000; Hine, 2015; Miller & Slater, 2000).

Gupta and Ferguson (1997: 12–15) point out that ethnography as a form depends on two kinds of writing: fieldnotes and the written ethnography. This literary form itself was isomorphic with the two spaces, the field site and home. They suggest, in a parallel way to the suggestions of Faubion and Marcus, that this separation, maybe at one time worked as a model: the fieldnotes happened abroad and the written ethnography happened at home. But as Fabian suggests, in order for this model to work, there was a denial of a shared life and shared world with one's informants. Now, denial is more difficult, because in a globalized world we do share a world with our informants. But maintaining the distinction between what constitutes fieldnotes and what is the written ethnography is now more fraught. To deal with these new contradictions, Fabian is suggesting in a more recent book that ethnography could become commentary and that a running blog might substitute for fieldnotes and ethnography.

The issues of ethnographic distance have been interesting and complex in my work with the Math Forum. But also my own identity has been caught up in this complexity. I began work with the Math Forum engaging in ethnographic evaluation of the original project to build out an online resource center and community for math education. As time went on I began to think about myself as the Math Forum's ethnographer. But in the world of the internet and where spaces of the field were not separated from the spaces of home, it became harder to think about whether Math Forum staff members were informants or collaborators. And it became more difficult to think about when I was a collaborator or ethnographer. One trope from traditional ethnography that has remained over this time is that I have remained an outsider and have not become a math educator. I have indeed entered into dialogue and contributed to thinking about social and group cognition in math education. I have also contributed to thinking about structure and agency in math education as well. So, in a sense, I have been part of the discourse in math education. Yet I continue to experience myself as an outsider to that larger community. Further, I do not consider myself part of the core Math Forum staff. And while there are some ways that I am a staff member, I am not employed by the Math Forum and I do not participate in their dialogue in the same way other staff members do. So there is a line where I remain, like the classic ethnographer, having one foot in the community and one foot out.

Hybrid Social Spaces

Another unintended consequence of the mapping the forum activity discussed above is that it began to contribute to my theoretical development. In the early days of internet research, the late 1990s and early 2000, there was a pretty strong tendency to see virtual and physical as a binary opposition. This was certainly true in the popular literature but also true, to a large extent, in the scholarly literature as well. Many people made comparisons between physical or face-to-face communities and virtual communities – there was a lot of excitement about the potential of virtual communities. The popular press had a lot of stories about the utopian potential of virtual communities. And many scholars, myself included, were writing about what we called virtual communities and were also beginning to think about virtual research methods as something separate from research methods used in physical contexts (Jones, 1998; Kollock & Smith, 2002; Rheingold, 2000; Wellman et al., 1996).

It's not surprising that the first wave of discourse, both scholarly and popular, was about the binary opposition between online and face-to-face worlds. First, this was the era of desktop computers. One logged on to the internet through a modem at a desktop computer. Getting to the virtual world was slow, which meant you had time to make a cup of coffee and a piece of toast in the physical world while you waited to get online. This enhanced the idea that there was a hard and fast separation between what Gibson (1984) would call cyberspace and "the meat." Further, these kinds of binaries are useful for thinking. They become the simple way we first understand things. Following Lakoff and Johnson (2003), the early explorations of the online world were through metaphor, such as the metaphors of space and co-presence and virtual selves. In these ways we began to ask how it is that the virtual world is similar to or different from our physical world. But as is always the case, these simple binaries are in fact too simple and obscure the fact that contemporary life, with the telegraph and telephone, radio and TV, was already thoroughly virtual, just not in the ways that computer and the internet made possible.

No sooner than scholars began to write about the binary opposition of virtual/physical, others began to question this binary and think about space, communities, and individuals as not either/or but both. In 2000 Daniel Miller and Don Slater published an important ethnography called *The Internet: An Ethnographic Approach*. In a way the book was a kind of sleight of hand, because it was an ethnography of Trinidad as much as it was an ethnography of the internet. Miller and Slater (2000: 1) state:

> Why should we do an ethnography of the Internet in Trinidad, or Trinidad on the Internet? Because – contrary to the first generation of Internet literature – the Internet is not a monolithic or placeless "cyberspace"; rather, it is numerous new technologies, used by diverse people, in diverse real-world locations.

They go on to discuss how Trinidad is really not just a location on an island in the Caribbean. It is rather a vast global diaspora, and to be Trinidadian is no longer about being in one particular place at a particular time. Rather, it is about being connected to a culture or set of cultures of people and part of an interconnected set of conversations, practices, and objects that are stitched together with the internet, among other technologies.

There are some important implications of the perspective that Miller and Slater take here. First is the realization that the physical world is itself mediated (and has been for a long time) and that these newer forms of mediation don't just create an online world but create a new world and one where group and identity get enacted in new ways that lead social actors to thinking about group and identity in new ways. Miller and Slater (2000: 8), drawing on the work of Latour, begin to map out a hybrid world where there are not binary distinctions between the internet and Trinidad as a place.

While Miller and Slater's conclusions are profoundly important, Boellstorff (2008) suggests that we cannot just dissolve the gap between the virtual and the actual. He suggests that the term "actual" is better for an opposition to virtual because both are quite real and necessary for human social life. Taking a somewhat broader perspective on "virtual" than many of the first generation of internet scholars, Boellstorff suggests that the virtual is connected to human imagination and reflects the potential of self, society, group. As such it is a critical part of bringing into being what actually is. In fact, there would be no future to the actual if it were not for this sense of the virtual. What is nice about Boellstorff's thinking is that it connects to anthropology's long preoccupation with Anderson's (1991) notion of the "imagined community" and it connects to the excitement that early internet researchers saw in the potential of online communication for advancing the self and the collective. Boellstorff warns that we do not want to rigidify the boundary between virtual and actual, but that the gap between the two is the creative space that makes life possible and interesting.[2]

While Boellstorff was working in Second Life and it was clearer that there was a stronger boundary between what we might call the virtual and the actual, he pointed out, importantly, that many of his informants felt that they could be more true to themselves online than they could offline. There was a way in which online felt more "real" to them and more "authentic," and in the actual world, they were forced to retreat from one another. This is an important and interesting insight. At the Math Forum informants made similar claims. First, Math Forum participants very much worked in a "hybrid" fashion. They worked together face to face and they also worked virtually online. The internet made it easier to share resources and ideas, but teachers always like to meet face to face. They used the virtual to imagine themselves, their students, and the ways they did math differently. And they were able, through that virtual work, to bring something new into being. The Math Forum staff was aware of these possibilities and worked hard to leverage the synergy one got from doing workshops or activities both online and face to face. They still run workshops in this way, where there may be some face-to-face participants and some online participants, and they may do preliminary or follow-up work online before and after a face-to-face workshop.

Once we got a little distance, as a research community, from the excitement of online communities and virtual communities and the ways they inspired (and still inspire) a kind of utopian ideal of human cohabitation, we began to realize we have always been hybrid. Or maybe as Miller, Slater, Boellstorff, Anderson, and others might say, we have always been virtual. In the internet age, email is the ideal example of how the internet contributed to this hybrid virtuality. Interestingly, email became so ubiquitous that few scholars even bother to think about the ways that it helped warp the social fabric. In the 1990s there was a little discussion about how companies were trying to use email to make workers more productive. They would encourage employees to contact others by email rather than getting up from their desks and going to another office. This suggested a sort of weird hybrid existence where you would email the person next to you. But of course this is a small example of the way that email both speeds up connection but also slows it down, and the way that email is integrated into people's work lives where they both see other regularly but interact virtually as well.

The Math Forum had lots of experience linking the online and the offline too. From the beginning it explored hybrid interactions and the ways they made it possible for people to develop deep connections with each other and carry on work over longer periods of time. We will discuss this more in the next chapter, but let me briefly lay out some of these Math Forum practices here.

The Math Forum began in the late 1990s, building a set of online resources and a community of teachers who worked with the Math Forum and helped it to build out those resources. Teachers were originally recruited through a set of summer workshops. The summer workshops were face to face, primarily on the campus of a small elite college, and lasted several days. Teachers applied to come to the workshops, and workshop participation was awarded to individuals that the Math Forum thought would make good candidates for being regular contributors to the website.

From the beginning the Math Forum made these workshops part of a larger hybrid social time/space world. There was always some pre-workshop work that teachers were encouraged to do. They were also encouraged to continue doing post workshop work online. This might happen in several different ways. Teachers would be encouraged to continue to develop their own personal webpages either on the Math Forum site or linked to the site. They were encouraged to contribute lessons and tools to the Math Forum. They were also encouraged to develop collaborations with other teachers from the workshop.

Sometimes teachers would begin a project from a workshop online, and then other teachers, who had not attended the workshop, would begin to collaborate with the teacher on the online tool or set of resources they were developing. In some cases these virtual collaborations would go on for quite a while before the teachers would meet up face to face at a professional meeting or at another Math Forum event. In the early days of the internet all of this hybrid interaction brought teachers together from different school districts and different parts of the country. It left teachers feeling empowered and as if they were part of a larger community that cared about mathematics and math education. And these hybrid activities were also supplemented by the Math Forum's regular online activities, the PoW, Ask Dr. Math, and a range of online discussion lists.

For me as an ethnographer, the experience working with the Math Forum helped me see early on that the hard binaries that people were reinforcing, in both the scholarly literature and the popular literature, between physical/virtual, online/offline, were a bit too rigid. The director of the Math Forum once said to me that the Math Forum staff was interested in the internet and new media technologies to the extent that they could enhance the ways that people were able to talk about math, do math together and share resources. He went on to say that community, such as it formed, was an effect of these practices, not the reason for doing these things. At the time I found this a powerful and profound statement. I still do. It was perhaps even more profound in the year 2000 as so many educational projects were driving to create online communities. For them, communities were the goal: for the Math Forum they were the effect of these hybrid practices.

Social Cognition and Communities of Practice

An important set of ideas from ethnographic theory for working with the Math Forum has been the ideas in social cognition, situated learning, and communities of practice. This work has been generated by ethnographers in anthropology, education, psychology, and a few other related fields. Central to these ideas is the Bourdieuian notion of practice and notions of how learning must be situated in a social context and the things that people do in that context.

A foundational concept is Bourdieu's concept of "habitus" and the implications that habitus has for all the social practices that exist within a social field. Habitus, for Bourdieu, is a structure. He framed it as a structure because even though we can see it as something internal to individuals, it is not a personality trait of that individual; it is something that is social and framed by the larger social context. Further, habitus is intersubjective: it is not only shared by individuals but is inculcated from one generation to the next largely through the institutions of family and schooling.

Bourdieu (1990: 53) states:

> The conditionings associated with a particular class of conditions of existence produce *habitus*, systems of durable, transposable dispositions, structured structures predisposed to function as structuring structures, that is, as principles which generate and organize practices and representations that can be objectively adapted to their outcomes without presupposing a conscious aiming at ends or an express mastery of the operations necessary in order to attain them.

For Bourdieu, habitus is developed through the education of the tastes as well as bodily disciplines. Individuals grow up in families, which themselves are embedded in larger social class (and class fractional) spaces. Attitudes in these groups about how one should hold one's body, eat food, appreciate (or not) art, read, and basically engage in everyday life structure an individual's orientation to the world. For example, in the case of bodily discipline, children who are taught to chew with their mouths closed not only will come to see that as the proper way to eat, but may be physically revolted by individuals who do not eat in this way.

While habitus becomes a structure that orients individuals toward their everyday practice, it also becomes a structuring mechanism for future activity. How I learned to think about time, space, and talk will have an influence on how I approach learning in school and what practices and tastes I think are appropriate in schooling or in other contexts. This is the real power of the notion of habitus, and the recent history of the sociology of education is filled with work that has developed the notion of habitus and how it works in different contexts (Lareau, 2011; MacLeod, 2008).

Lave and Wenger (1991) took up Bourdieu's notion of habitus in their small and influential book *Situated Learning: Legitimate Peripheral Participation*. In this book they situate habitus in a larger structure of activity that they referred to as the "community of practice." Looking at small apprenticeship groups, such as midwives, tailors, and butchers, they wanted to think how learning worked itself out in practice, and what implications this might have for the field of education more broadly. They drew on Bourdieu's notion of habitus and field to think about how learning in these groups was situated in the practices they engaged in. But further they thought about not only how communities develop but how their knowledge grows through new apprentices who start out as novices and advance the practice of the group over time, and finally, how this process is tied up with individual and group identity. These ideas are at the core of the social cognition work that came after.

There are related sets of ideas to situated practice that many scholars have brought together. Vygotskyian psychology and activity theory attempt to situate learning and cognition within a complete activity system that would include everything from the macrostructural model of production to the particular contexts of tools, language, and relationships within which a particular practice was occurring. While there might be some intellectual tension between activity theory that grew out of Marxist theory and Bourdieu's practice theory that could be seen as more Weberian, there were several efforts to bring these two bodies of theory together fruitfully (Chaiklin & Lave, 1996; Kirschner & Whitson, 1997; Rogoff & Lave, 1984).

In a similar vein, the linguistic research of Mikael Bakhtin (1981) has deeply influenced the research in this area. Bakhtin's deep understanding that language and discourse are social and that the forms of discourse are deeply related to forms of thought and have a life of their own outside the individual are an important set of ideas for social cognition. In cognitive anthropology, there has been an emphasis on discourse and the narrative structuring of thought, and then the relationship of those narratives to issues of personhood and identity, both individual and collective (D'Andrade, 1995; Holland et al., 1998; Strauss & Quinn, 1997; Wertsch, 2002).

In this work there is a dynamic interrelationship between the practices that people engage in, the discourse around those practices and activities, and the ways that people think about what they are doing and create knowledge. These processes are both individual and collective. They are also dialectical in a general sense that one moves from practices to discourse to thinking, back and forth and around the circle. Each helps develop the other. Knowledge generated in this process is intersubjective. We could think of this process as learning, knowledge production, or intersubjective meaning making (Bruner, 1996; Suthers, 2006). Wertsch (2008), discussing the problems of collective memory, suggests that memory has a narrative structure and that members of a community might each have pieces of that narrative. A whole narrative is constructed through these parts. We can think of knowledge in the same way. It is intersubjectively generated, and different individuals hold different parts of the narrative. This is why working groups,

Thinking

Discourse ←——————→ Practices

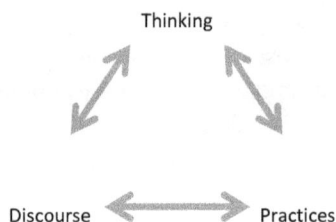

Figure 2.1 The Math Forum Dialectic

when they get back together after having worked on a problem previously, need to reconstruct where they have been and what they were thinking. It also means that knowledge is somewhat fluid and each effort of intersubjective meaning making might yield new and different insights.

As I mentioned in Chapter 1, the staff at the Math Forum, from the early days of the Math Forum's existence, articulated a similar kind of model for their vision of mathematics education. Math, for the Math Forum, is part of everybody's everyday life. And many of the problems that the Math Forum developed for their website in the PoW section came from their own everyday lives as they interacted with other people. Math is something that people do; it is a regular practice. As their own community of practice, the Math Forum staff have always done math together. When they get together with teachers, either in face-to-face or online workshops, they always do math together. Similarly, doing math together requires talking about math together. It is always an intersubjective activity. The learning for the Math Forum is social in the ways we talk about what we figured out, where we were stuck, or the new insights we had. And thinking/knowledge production is part of that process of problem solving and discussing problem solving. I have referred to this necessary connection between the practice of problem solving, the conversation around the math that one is doing, and the development of new ideas or thought as the Math Forum dialectic (Figure 2.1).

Critical to the Math Forum view of learning and mathematics education are new forms of social arrangement that prioritize communicative interactions as one of the key features of the way the social space works. And thus, the internet is a real asset to thinking and learning. If conversation and interaction lead to deeper thinking, it also leads to the formation of a discourse community. Peirce (1931) pointed out that all thought is dialogic and that even when we think alone, there is always an interlocutor and an imagined conversation of which are always part. Bakhtin (1981) made a similar argument in his work *The Dialogic Imagination*.

In math education, Anna Sfard (2008) brings together the work of scholars such as Dewey (1938), Vygotsky (1986), and Wittgenstein (1953) and claims that communication and cognition are flip sides of the same coin. She suggests that our traditional ideas about knowledge acquisition are incorrect. Knowledge and what we call learning are communicative acts and necessarily social. To capture this insight, she invents the term "commognition" to underscore that communication and cognition are social and intersubjective (Sfard, 2001). To Sfard, mathematical development involves being assimilated to a new discourse, one that has conceptual objects for which the student has no former understanding. In this way learning is similar to scholars or scientists developing new knowledge. Using the semiotic notion of signifier/signified she points out that students have no choice but to use a signifier with which they are not yet familiar. The way they develop the mental objects or conceptual understanding is akin to the ways that Lakoff and Johnson (2003) talk about how metaphor is used to expand knowledge in general (Sfard, 2001).

The problem of learning mathematics then is related for students, teachers, and mathematicians. They must find themselves in a discourse community where problem solving and mathematical practice is part of the norms of that social group. Being a member of that discourse community leads in a dialectical way to more conversation and more thought and deeper forms of knowledge and understanding. Lave and Wenger (1991), for example, note that it is impossible to distinguish the learning from the context within which the learning takes place.

While not an ethnographer, Sfard's ideas are similar to those of ethnographers working with ideas of social cognition and practice theory. And all of this academic work bears a strong resemblance to the ideas that the Math Forum staff has developed in practice. As stated above, for Math Forum staff the key value of the internet was that it not only created more opportunities for mathematical conversations; it increased the ways those conversations could be had, and then later reused. And in the same way, conversation and activity led to the creation of a discourse community where people could develop attachment and belonging. The new media tools helped facilitate the development of the Math Forum discourse community and made it possible for individuals to deepen their activity and conversation and support identity shifts as well as shifts in the social imagination of the group.

Ethnography has been an invaluable tool for exploring this weave at the individual and social level. It has also been critical to the formulation of theory as to what is going on. As Herzfeld has suggested, ethnography is a form of theoretical practice.

The anthropological view of ethnography is that it is more of a stance than a method. It is a way of engaging with other people, and involves an ethics toward the others with whom one engages. The goal of that engagement is to better understand the experience of the other, the way that the other engages in their own practices, and how they understand the problems that they must try to make sense of. Ethnographers do this through theoretical engagement. We make sense of our experience with the other through theory.[3] By engaging with the Math Forum I have learned to appreciate their understanding of learning, mathematical knowledge, the power of the internet, and the problems of trying to keep a small internet organization going in the twenty-first century.

The Rise of the Digital

Recent discussion in the scholarly literature has shifted from the distinction of online/offline or virtual/physical to the digital. We now pose the digital to the analog whether we are talking about technologies, communities, or cultures (Coleman, 2010; Horst & Miller, 2012; Jackson, 2012). Coleman (2010), in a review article on the digital in anthropology, suggests that ethnographic work can be thought of as moving into three areas, what she calls the "politics of cultural representation," "digital media vernaculars," and "the prosaics of digital media." Coleman's first area is similar to the work that Miller and Slater did in Trinidad about how a group of people imagine themselves and connect with each other through internet technology. Coleman's other areas of research refer to the growing body of studies that focus on things such as the ways that digital media are creating new contexts and forms of interaction, political and economic activity, as well as new forms of language and cultural experience.

Boellstorff (2008: 18) was originally critical of the use of the term "digital." His critique suggests that digital is almost a synonym for electronic and that so much of contemporary life is digital. The term becomes very broad and does not have much analytical purchase in his view. Later, while still critical of the notion of the digital, he suggests that digital anthropology could be the space where we sharpen our understanding of the relationship between the "virtual and the actual" (Boellstorff, 2012: 56–57). Echoing his thinking about the gap between the virtual and the actual in *Coming of Age in Second Life*, Boellstorff continues to think about the constitutive role that the gap plays in this later work. Boellstorff suggests that new technologies, as we move from web 1.0 to 2.0, are not blurring the virtual and the actual, that digital is not about blurring. It may very interestingly have complicated the movement between these spheres, but without the gap there is not imagination or becoming. This is a critical point for thinking about the Math Forum as an internet resource center.

Emphasizing the analytic value of the notion of the digital, Miller and Horst (2012) suggest that the digital can be an important way of thinking about transformation and change in contemporary cultures. Through a dialectical analysis, they suggest that like the evolution of money in contemporary societies, the digital moves us from the particular to the more abstract. As money has become more abstract, it moves further away from direct exchange relationships between people meeting in a marketplace and toward a more universal and impersonal form of exchange. So too the digital moves us increasingly toward more abstracted social relations that can be recorded and reused in a multitude of situations. Miller and Horst point out in Hegelian fashion that as these tensions create a conflict and individuals attempt to reconcile the contradiction. They suggest that in the case of money, people attempt to deal with the alienation from genuine human relationships with commodities and things. In the same way, the digital introduces us to a world of digital commodities where this tension can be partially reconciled. They further suggest that with the digital, the ability to produce and reuse every kind of object introduces an interesting new potential for a "decommodification" of objects to fill in the space between human connection and alienation (Miller & Horst, 2012: 7). A dialectical perspective leads them to remind us that while there are pessimistic views of our digital futures as well as utopian, these tensions cannot be overcome simply.

Miller and Horst's perspective leads to two further implications. First, there is a way in which they are theorizing a kind of gap too. There are parallels to Boellstorff's notion of the gap between the virtual and the actual, with their gap looking at the way the digital exacerbates the problems of reification, while the digital at the same time is used to attempt to reconcile that tension. Second, this dialectical tension at the heart of the digital is precisely the tension the Math Forum seeks to reconcile. It does this by trying to use the internet to increase human connection and it attempts to use digital products, people posting problems, lessons, good answers to math questions, etc., to improve people's access to good math resources, all the while exploring and weaving across the gap between the digital and the actual. We will discuss all of this in later chapters.

If a globalized, interconnected world raises questions about the separation of fieldnotes from finished ethnography and encourages a whole rethinking of the ethnographic project, John L. Jackson (2012: 483) asks, "What kind of writing does the potential digitalization of culture demand and afford?" Jackson begins the article by talking about how the documentary filmmaker Marlon Riggs and the anthropologist Barbara Meyerhoff each filmed their last days of life in a kind of digital autoethnography. Jackson questions this radical digital reflexivity in terms of the contradictory space opened by *Writing Culture* and its emphasis on the aestheticization of ethnographic writing. By opening this discourse, Clifford and Marcus (1986) helped to legitimate at least the consideration of the aesthetic in ethnography.

Jackson then goes on to tell a story about a group of people he has researched, the African Hebrew Israelites of Jerusalem (AHIJ), and their selling of a piece of equipment that is designed to take individual frames out of a video, without disturbing the flow of that video. The result is the "same material" in a shorter time that allows advertisers to slip in more commercials. Jackson later tells us about how the AHIJ are experts in digital products because so much of their work as a religious community is tied up in both the selling of digital artifacts training followers in the belief, and the creation of documentaries about their founders. The two stories frame some of the ways that the "digital" is affecting all of our lives, whether we do "digital ethnography" or not. Space and time are altered in different ways by the digital, and, as indicated in some of the other examples above, we are all part of this process. But importantly, many people are now narrating their lives more self-consciously, and we ethnographers are no longer the "architects" of that narration but rather entering into the process with others in a complementary fashion. This refashioning of the relationship is also interesting because for Jackson, the digital tends to be produced in a commodity form, whether that form is for sale or freely distributed on Facebook, for example.

Ethnography and History

There has always been an important relationship between ethnography and history. Cultures unfold through periods of time. And importantly, cultures find themselves tied up in economic and political configurations that have certain kinds of tensions built into them and have different lines of force and pressure within them. In that way, the evolution of a culture cannot be predicted, but can often be clearly explained in hindsight. This is where having almost two decades of work with the Math Forum is a real asset. These historical trends will be explored in the next two chapters, but the groundwork can be briefly laid out here.

When the Math Forum began in the late 1990s, the NSF thought it could be a model for other online educational resource centers and online communities that could support teachers, students, and families. In the early days their focus was replication, scaling, and sustainability. How could a site such as the Math Forum be cloned? And how could these educational communities be scaled up to handle a nation of users who might want to have access to their sites and interact with each other? Finally, how could this activity be sustained without continued input of federal funds?

These were huge questions that still have no answer. But they were posed in a spirit of optimism because during the dot-com boom everything seemed possible. Of course, the boom was not perceived as a bubble; it was thought of as a genuine change in the way the world worked, thanks to the internet. When the dot-com bust came, only then was the boom seen as a bubble.

Likewise over this period of time we went from web 1.0 to 2.0 along with the expansion of the technological infrastructure and the increased complexity of programming for the web. These changes created an interesting irony. Web 2.0 is seen as the interactive web, but it is an interactive space that has become more commodified and complex, so we interact as consumers. But web 1.0, while more fixed to desktop computers, was in fact more interactive where anyone could write simple HTML code and create their own web pages without the need of any templates or support. It was also the era where a less fettered communication flourished in places such as the Usenet and other unmoderated discussions.

An important part of the Math Forum story is about how it was born of the utopian spirit of the early web and the utopian views of the Internet in the 1990s. And then how the Math Forum managed to survive the changes from the 1990s to the present. On the one hand, it is a remarkable that the organization has survived and done well. On the other hand, we will see how it could have been even more remarkable than it has managed to become, had conditions and influences been different.

Notes

[1] While he was working at Microsoft Research, Smith developed a platform called Netscan that allowed you to see several things about discussion lists. It allowed for the mapping of the list in relationship to other lists. It also allowed you to see who was the biggest poster and what kinds of posts they made, whether to initiate

a thread or respond to other threads. More recently he is part of Connected Action and has been one of the key contributors to the NodeXL project, www.connectedaction.net/marc-smith/.

[2] Boellstorff's thinking about the virtual and the actual is similar to Spinoza, Flores, and Dreyfus's thinking about entrepreneurship in *Disclosing New Worlds, Entrepreneurship, Democratic Action and the Cultivation of Solidarity*. Spinoza, Flores, and Dreyfus point out that entrepreneurs also bring into being new worlds that cannot even be thought of in the present world, and they do this by working across the gap they refer to as disclosure.

[3] Herzfeld points out that informants make sense of their worlds through theoretical practice as well. The difference is that our theories are ones that have been developed through scholarship. I don't know that it makes our theories better than the everyday theories of our informants, but it makes them different and part of a different kind of conversation. In my work with the Math Forum, my greatest insights came from the theoretical practice of my informants, which often I could connect to academic theory.

3   History of the Math Forum

Introduction

In Chapter 2, we concluded with the importance of the relationship between ethnography and history. Ethnographies are situated in local cultural contexts and look at the lives of the people in those contexts. But local contexts are themselves always situated in larger structural frames that are regional, national, and global. Further, cultures unfold over time. They are transformed by historical circumstance, but more importantly they become what they are through time. Our history of the Math Forum will look at the unfolding of this culture and institution at several levels. The chapter will focus primarily on the local history of the Math Forum, but by necessity will make some connections to the larger structural context. Chapter 4 will look more exclusively at that larger structural context to think about ways that it shaped the Math Forum.

I've written other histories of the Math Forum at different points in time. In our early research, my colleague Ann Renninger and I mapped out an early history of the Math Forum (Renninger & Shumar, 2002; 2004). In that work we were mostly interested in the development of the nascent culture that the Math Forum was becoming and the impact it was having on teachers (and students and others too) as the Math Forum began to grow. "Communities, Texts and Consciousness" was probably the most developed "history" of the Math Forum (Shumar, 2009). The narrative that I constructed at that point drew on these three analytical concepts to talk about what the Math Forum was, a hybrid community of participants, the traces of conversations, problem solving, discussions that became reused text, and the consciousness of a particular way of seeing everyday mathematics in the world. That model was a powerful one for thinking about the Math Forum. It obviously echoed the dialectic of practices/discourse/thinking that was laid out in the previous chapter. I will draw on some of the insights in that chapter here.

There are different ways to frame the history of the Math Forum and here I will attempt to bring three frames together. I will talk about the history of the Math Forum as the history of the places where the staff of the Math Forum was located. These locations are a good way to periodize the development of the Math Forum. They also make connections to people involved, the ways the culture was developing, and the larger structural pressures that the Math Forum experiences. The second frame is the history of the projects (largely NSF funded) in which the Math Forum has been involved. The research projects that the Math Forum has been part of has allowed them to build out their online services and resources. These projects have also shaped the experiences of the Math Forum staff with site participants, primarily teachers, and has allowed them to think about how they can support teachers in the development of their own mathematical thinking. And then as teachers increase their ability to reflect on how mathematical knowledge is built, the staff has thought about the ways this knowledge can inform teachers' teaching as the teachers think about how students think about math. Between the discussion of locations and projects, the third frame I will use in this chapter is a story. The narrative is meant to give insight and detail as to how the culture of the Math Forum developed and got expressed in practice. The story is of the summer workshops in San Diego and Philadelphia in 1997, and the point at which I really began to see how the Math Forum operated. The Math Forum, as an organization, was only a few years old at that point. And the summer workshops in 1997 were built on workshops the Math Forum had done in 1994 and 1995 before I was part of the Math Forum. These early workshops not only give us insight into the Math Forum culture, but they are the model on which all future work will be based in one way or another.

Locations

The Math Forum has really had five main homes. These are different physical spaces where the organization developed. The first two were on the same small college campus in the mid-Atlantic. So they have had four institutional homes. Each of these locations corresponded to different phases of the Math Forum's development. So it is worth talking about each for a bit, and talking about the issues the Math Forum was facing at that time. We cannot do that without making some connections to the material in Chapter 4. But here we will not develop those ideas very much but rather just point to those larger structural issues.

The Fish Bowl

The Math Forum was born at Swarthmore College, a small elite liberal arts college in the Philadelphia suburbs. It began before the web and was originally the geometry forum, a discussion list on the Usenet. It was the idea of a math faculty member and a math student at the college. Both are still active members of the Math Forum. The pre-web discussion lists were all text, and so the Math Forum was born in an environment where talking about math, talking about how problem solving could be done, and talking about pedagogical issues were the main activities.[1] Around when the web was born, the NSF asked the folks at the college if they would consider creating a forum for all of math, not just geometry. They received a one-year "proof of concept" grant and then after that a three-year grant to develop the Math Forum. In the early days of the Math Forum they had one office at Swarthmore College. They called it the "fishbowl," as it had a lot of glass and there were about half a dozen or more people working in this office at any time. The Math Forum in those days was as much an internet organization as a math organization. It started a library of resources, which is now called the Math Forum's Internet Mathematics Library, but originally it had a diverse array of internet resources and links. In the early web before Google, having an organized repository of links and resources was a real asset.

Much of the focus of this early Math Forum community, which worked so closely together in the fishbowl, was to develop the online services Ask Dr. Math and the Problem of the Week (PoW). Ask Dr. Math is a question-and-answer service for mathematics. The service began with live mentors who served as "math doctors." They would answer people's math questions. There was a "tenuring process" where volunteers to be math doctors went from apprentices themselves to full mentors. Their replies to math questions were reviewed by existing tenured doctors for quality and consistency. After an apprentice had answered a number of math questions well, they became a "tenured" doctor and then they had direct send – they could send out their answers to questions directly. They were also then able to mentor other apprentice doctors. The PoW was a nonroutine weekly challenge problem. The problem was posted at the beginning of the week. Answers to the problem were then responded to by volunteer mentors. Students who answered these problems were encouraged to clearly show their work, communicate their thinking, and think more about each problem. Every week some of the best answers would be displayed on the website for all to see. Many teachers used this service as an adjunct to classroom activity. The PoW began as the geometry PoW, but different PoWs developed later.

These early services not only provided resources to people but were interactive. Further the staff had the early internet idea of taking past interactions on the website and repurposing them as resources to be used by other. Much of what becomes the Internet Library are past interactions, questions, answers, problems, and solutions that can then be mounted as resources for others to use. I discuss this idea of the reuse of past inter- actions in detail elsewhere (Shumar, 2009). This attitude toward text was important because interactions were seen as a resource, not just fleeting moments. If someone asked a good math question and there was a good answer to that question, provided by either a staff member or a volunteer – that interaction was "capital" and should be carefully repurposed so that it could be used to its best advantage.

It is also in this fishbowl environment that the heart of what I would call the Math Forum culture comes together out of the shared practices of this group of people who are working so closely together. They have a set of online services and resources that they are working on. The most important are Ask Dr. Math and the PoW services, but there are discussion groups and other activities on the website. Then there was a series of partnerships that the Math Forum fostered. Some of these partnerships were individuals, such as math professors or teachers who shared the Math Forum's vision, and others were institutional partners, such as the Mathematics Association of America (MAA), a school, or an NSF project. Obviously, the institutional partners involved individuals too. The Math Forum's attitude toward these partners was similar to their attitude toward the internet in general. The thinking was that these new technologies allowed for collaborations of a much more closely coordinated type: what the group needed to do was to stay in touch with people and keep brain- storming about projects or activities that they could do together. Some of this was thinking about projects for future funding, but most of it was just thinking about ways to work with students and/or teachers in order to do more math together and to talk more about mathematics. To my mind, while many others who were involved in the early internet culture were mystified by the power of the internet, the Math Forum staff was amazingly clearheaded about their situation. They neither glorified the technology nor did they ignore their life off-screen for their life onscreen. They were always aware that the technology was a means to improve communication and to interact with people across distance and time. They focused on the communication, which is what tended to build the community.

The College House

Eventually it became clear that the activities of the Math Forum were too big for the space they were occupying. They were moved to a small house at the southern end of the campus. The house was still a small space, but it did provide more room for the staff to work. It was a detached house where they clearly had their own space and were not occupying a room in the math department. In this house the summer workshops in 1997 were planned.

In this space the work to develop community members continued. Several staff members at this time focused on reaching out to collaborators and to attempting to brainstorm possible ways to collaborate and develop resources. While the Math Forum staff was clear that the internet could have big implications for education in general and math education in particular, they were much like a young entrepreneurial company at this point. They were looking for new partnerships and new ideas in order to develop new products and services. While they were not planning on being a money-making enterprise, they did have the goal of being self-sustaining. The faith that everyone had in the internet at this point was that it would be possible to figure out how to create a self-sustaining organization. From the beginning the NSF project plan was to wean the Math Forum of federal money and to also spin them off from the college. The thought was they would find their place in the internet economy and be able to continue to provide high-quality educational services.

Another critical issue that the Math Forum faced in these early days was how to scale activities, especially with the two main services, Ask Dr. Math and the PoW. Even though there was a volunteer workforce working on these services, there was an enormous amount of work to be done, beyond what the volunteers could do. And yet the volunteers needed to be managed. The problem was especially acute in the case of Ask Dr. Math, where hundreds of math questions came in per day. While they tried to answer as many questions as they could, the head staff person for Ask Dr. Math focused on archiving the best questions and the best answers to questions. The Math Forum realized that if they could automate most of Ask Dr. Math, where people could quickly and easily find good answers to their questions, this would address the scaling issue. The math doctors then could focus on more unique questions, as the bulk of questions had answers in the archive. The issue of scaling was a bit more complex for the PoW services. This was because the problems were non-routine challenge problems; the point was for kids to get a reply from a mentor and be encouraged to revise their work. Even if they had the right answer, they would be encouraged to have additional thoughts. But also since the PoW was done by kids in classes either for extra credit or as part of a classroom challenge, the number of respondents, while large, was not as great as the Ask Dr. Math responses. The Math Forum did the best it could with staff members and the volunteer workforce. The PoW situation would change when Math Forum went to Drexel University, but for a long time it was a free and open public service.

Executive Office Building

Anticipating the "spin-off" and that the Math Forum would be acquired by some other internet education company or organization, the staff moved operations from the small building on the college's campus to the one office building in the suburban town where they were located. The office was a large open floor where many cubicles could be installed. There were also a few offices around the exterior walls. Even with cubicles there was a large open space where people could kick a soccer ball, play with other toys, and think. The Math Forum began to resemble what we have seen in other internet companies where the play space allows for creativity and is closely linked to the workspace.

Shortly after the Math Forum moved into this larger space, it was acquired by WebCT in April 2000. Technically, this was after the dot-com bubble began to burst, but of course at that moment in time, there was still a lot of optimism about the internet and not everyone saw the bubble as a bubble. WebCT had grown because of its main product, a course management system (CMS) or learning management system (LMS) that allowed faculty to put up recorded lectures, readings, and other material; have asynchronous discussions; and so on. The system could be used either for fully online courses or as an adjunct to a face-to-face class. The primary market for CMSs were universities and university faculty. And the WebCT business continued to grow beyond the dot-com crash until it was bought by Blackboard in 2006.

WebCT had cornered the early market for CMSs. But its executives, like a lot of internet-based services, began to think about content. Their idea was to use the principle of the internet and get users to help them develop communities around subject areas. The thought was that this would be another reason a faculty member would want to use WebCT. Not only was it a platform for online learning, but there would be lots of resources and colleagues in one's subject area that a person could draw on to help make their own courses richer and more interesting for their students. It was a progressive and expansive idea. The Math Forum did not neatly fit WebCT's model, as much of their resources were geared toward K-12 education. But as Math Forum was the main source of online math education material, the thinking was that it could bridge the gap and find ways both to expand in the direction of K-12 and to have Math Forum resources available for university faculty.

Part of the arrangement with WebCT was that the director of the Math Forum would become an executive at WebCT and spend much of his time in Boston where WebCT was headquartered. This raised an interesting leadership issue for the Math Forum. Like a lot of small organizations, the Math Forum's director is a charismatic and creative person. He has a strong knowledge of math education and was a former teacher. As an early internet adopter, his understanding of programming and technology was strong too. Also with a background in conflict resolution, he had excellent management skills. Anyone who has known him understands that he is a unique leader. These skills are hard to find in one person; nevertheless, the Math Forum staff began to talk about "cloning the director." This was a way of saying that for scaling and sustainability they needed more strong and unique leadership. While the Math Forum never developed another strong, singular leader, they did hire and support a number of people in their leadership development. Among others, the head of the programming group was a strong leader during this period of time, as were some of the other original staff members of the Math Forum.

Only a few short months after WebCT acquired the Math Forum, WebCT decided to spin them off. While WebCT survived the dot-com crash, it felt it needed to move away from the progressive vision of content communities and focus on its core business of CMS. There were other competitors and the business had to become very focused. WebCT planned for a termination date and set up severance packages for the staff member. Math Forum brought on a colleague who had expertise in putting together groups who needed financing in the hopes that he could help the Math Forum find a new home. Just when it looked as if there would be no future for the Math Forum, a deal was brokered with Drexel University and the Math Forum joined Drexel in June 2001.

Drexel University

The Math Forum was at Drexel from June 2001 until June 2015. It has been the longest period of a stable home for the Math Forum. The Math Forum moved from a liberal arts culture and the early utopian views of the internet to a research university that was focused on the financial bottom line. Immediately, Drexel asked to reduce the size of the Math Forum staff so that their budget needs would be smaller. Further, the university wanted a plan as to how the Math Forum could monetize their services. Finally, they were charged to think about ways they could provide service to the undergraduate population at Drexel. All of these new requirements were about trying to make the Math Forum self-sufficient. The university was aware that this was a goal and not a reality. The view was that this kind of positive pressure on the Math Forum was good; it would make them creative and thrifty. Yet the university did commit to meeting any shortfall that occurred.

Monetizing services primarily meant turning the PoW from a free service to a pay service. The PoW was monetized along a couple of different models. Schools and school districts could pay for PoW services. Individual teachers could also buy their own accounts if their school did not have an account. A basic PoW service was available for free, but it did not give people access to the PoW archives nor did it allow them to have their students mentored. Thus, the free service was a limited service. Understandably, some mentors who had worked as volunteers in the PoW were alarmed at the transition to a pay service. But the Math Forum attempted to work with them and make sure they were comfortable with the transition. Monetizing the PoW services was done without much support from the university. The staff was left to think on their own about how services could be marketed to a larger constituency. Further, the staff ended up doing a lot of more traditional teacher development workshops for teachers in the Philadelphia area. This drive to have the Math Forum pay for itself was a smaller vision and often had the Math Forum working in ways that did not make best use of their skills or potential.

During its time at Drexel, the Math Forum had a number of its main research projects that I will describe in the "projects" section of this chapter. These projects were important for advancing thinking about the online learning of mathematics and the development of teacher communities. While this was never a stated goal of Drexel University, the Math Forum became a research incubator at the university. The Math Forum was a key player in NSF's digital library program and they contributed to thinking about cyberlearning in that community. The university did incentivize the funded research work by giving the Math Forum a generous percentage of their overheads from these grants to contribute to their bottom line.

The research center idea could have been an important way that the Math Forum fit into the larger university. But in the last couple of years of their time at Drexel, the university began to make plans for a responsibility center management (RCM) budgeting model. Built on the neoliberal idea that institutions are more rational if they are based on market models, RCM, in all its guises, sees the institution as serving a primary market, "students as customers," and then attempts to "rationalize production" based on that metaphor. In the RCM model, one is either serving the customer directly or providing services to those who are serving the customer. In this model, Math Forum was not providing enough direct services to university students or to other units serving students to justify their budget. After several efforts by many groups, it was determined that the only way to keep the Math Forum at the university was to change them into something they were not. As a result, the university and the Math Forum parted company in the summer of 2015.

NCTM

As of the writing of this book, the Math Forum has just finished its transition to the National Council of the Teachers of Mathematics. They have always had a good and collaborative relationship with NCTM and regularly had Math Forum booths at the national conference and some regional conferences. It appears that the Math Forum@NCTM is a good fit. Hopefully, Math Forum will have more space to concentrate on its core activities of supporting teachers and students of mathematics. And it should be able to continue to collaborate with researchers and develop new areas of research into cyberlearning in math education.

San Diego and Philadelphia

In the summer of 1997 the Math Forum held summer workshops in Philadelphia and San Diego. These workshops were part of the NSF's recently developed Urban Systemic Initiative (USI). The USIs were to support the needs of a nation moving into the twenty-first century where prolonged and coordinated efforts were needed for improving teaching and learning in science, math, and technology. These initiatives brought large-scale and multipronged coordinated efforts to have an impact on the education in particular urban locations. As part of the USI, the Philadelphia and San Diego school districts had decided to focus on math as an area to improve teaching and learning.

I traveled to San Diego to observe the two-week workshop that the Math Forum was doing with a group of teachers in the San Diego school district. As part of the USI, the Math Forum was invited to come to San Diego to conduct one of their summer workshop programs that taught teachers how to build basic websites. The workshop also taught how to then use those websites to improve their teaching by putting lessons online, providing resources, increasing conversations with other teachers of mathematics about issues that the teacher was facing, or all of the above.

The San Diego workshop was a replication of the original summer workshops that the Math Forum had done at their home, Swarthmore College, with teachers who were invited to come and participate. These early workshops would establish a model for how the Math Forum worked with teachers. While there have always been many different constituents who use the Math Forum site, and even be part of the community (e.g., parents, students, university professors, people who just like math), teachers have been the core of the active community and were the ones who tended to contribute the most resources to the site. This ethnography focuses more on teachers, because I see them as the core of the community. But this ethnography very much involves teachers' interactions with students too. While I did not attend the earliest summer workshops in 1994 and 1995 at the Math Forum, the workshop in San Diego followed a similar pattern to these earlier workshops.

The high school where the workshop was held was in the inner city of San Diego in a poorer neighborhood. I remember thinking as a lifelong East Coast resident that it was hard for me to see economic deprivation. All the houses were detached with palm trees in the yards. The neighborhood looked quite nice to me. But if you looked more closely, you could see that the cars were older and some were dilapidated. Yards and houses were a bit run down too. This was a poorer neighborhood, but it did look quite intact. The school was a typical West Coast campus with several smaller one-story buildings that were detached. Doors led to the outside and there were no interior hallways. The computer lab where the workshop was held was quite nice and we all were very comfortable. The weather was typical southern California, warm with low humidity. It was a welcome break from heat and humidity for us East Coasters.

The workshop was made up of about eighteen participants. The participants came from five elementary schools, two middle schools, and a high school as well as a community school and a teacher working across schools. In the workshop the Math Forum had several different agendas and activities associated with those goals. First, and most excitingly for the teachers, Math Forum staff taught them to build websites. In 1997, basic HTML coding was fairly simple and one could find an HTML codebook on the web. This was well before coding for the web became complex, and well before there were stock templates available for people to use to build websites. HTML, while fairly simple, was esoteric and looked complex and impenetrable to people who were not familiar with it. In fact, computing in general was a new skill for people beginning to use the web. I remember talking to a staff member at a student workshop where we noticed that college students who had more computer skills tended to use keystrokes more often to do things like cut and paste rather than clicking the mouse and using a drop down menu. I also remember seeing several teachers who were quite baffled by the whole representation system of micro computing. The idea of a mouse as a pointer and the ability to click things on the screen was a lot to grasp. Many teachers had worked with computers before and were not so far back in terms of their basic learning. Yet there was a range of skills among teachers, and very few of them had any sense of how the web worked and how to code in HTML.

Web page building in the workshop followed a simple pattern. Teachers were taught a few basic codes to set up a page, change fonts to bold and italics, create tables, and so on. Then they also downloaded some software that would let them copy images, for example, and upload them to their sites. These lessons in web page building were extremely empowering for many teachers. They went from feeling that computing and the internet were a high-tech mystery to feeling like they had learned a whole lot about how to be tech savvy. While in reality the lessons were pretty simple and the knowledge gain was modest, the experience for participants was much bigger than that.

Web page building was interspersed with math activities. The workshops were not about teaching math to teachers, even if some did learn some things in the math exercises; rather, the math activities were about doing math together. *Central to Math Forum belief is that when people work together on math education topics, they should do some math together.* Math activities were both opportunities to work on math and talk about math. But they were also opportunities to see what people had done with math and the internet. And so participants were about to see how other teachers and Math Forum staff had put lessons on the web and the ways the internet is used to support student interest and ability to do math. Finally, the goal of the workshop was to support teacher projects. Teachers were encouraged to develop projects that they could start during the workshop but then continue to work on during the year.

The teachers in the workshop were supported by a group of online mentors who were virtual participants in this face-to-face workshop. They were individuals that people could reach out to for ideas and help in particular areas of expertise. The summer workshop was followed up with a series of weekend online and face-to-face workshops. These workshops were designed to work on projects, explore Math Forum resources, such as the PoW, and work further on the value of the web for math education. The idea was to provide some sustained support and encourage teachers to use these projects to do something that was important to them.

The workshop in Philadelphia followed the same model as the workshop in San Diego. It was held at an inner city high school in North Philadelphia. There were about twenty participants, twelve teaching in elementary schools, three in middle schools, and five in high schools. Unlike the school and the weather in San Diego, which felt relatively easy, the situation in Philadelphia felt much harder. First it was hard to get into the building, which was closed for the summer. There was no air conditioning and the weather was hot and humid. At first, the workshop participants thought it would not be possible to do the workshop under these conditions, but then the Math Forum sprang into action.

The staff began to discuss with each other how they might work on the problem of the workshop location. They then figured out who had a spare air conditioner and some box fans. They brought these things into the location and began creating an environment that was reasonably comfortable. It was still hot, but it became possible to work and the workshop continued for the two weeks without further problem. The teachers were impressed with this kind of problem solving. It is clear that they had never seen people deal this way with a problem in a school. And it was also clear that this kind of problem solving was the same as the way the Math Forum staff dealt with mathematics. Problems were there to be solved and that is part of what people do. My colleague Ann Renninger and I were impressed with the problem-solving orientation and how it generalized from mathematics to other kinds of basic problems (Renninger & Shumar, 2002; Renninger & Shumar, 2004). Ann taught me how to focus on the problem solving that people were doing and the way that the Math Forum staff members were creative and resilient. In another workshop, a staff member famously bought modems and ethernet cable from a Radio Shack in order to string wires outside the schoolroom windows so that everyone could have a connection to the internet in the workshop. Not only did Math Forum problem solving tend to generalize to other areas of daily life, but this attitude informed how the Math Forum viewed math. Math was just another of those everyday things that people had to problem solve about.

Once the solution to the temperature was sorted out at the Philly workshop, it ran similarly to the San Diego workshop. Because Philadelphia was home base for us, it was a bit easier in some ways to do the workshop even with the problems it brought. The follow-up with the Philadelphia group was similar too. There was a set of weekend workshops during the academic year where people could continue to pursue their projects.

Projects

In this section of the chapter I will talk about several projects that the Math Forum was either part of or had initiated. The projects, while interesting in themselves, give us a sense of the different periods of the Math Forum's life as an online community. This section is divided into four parts. The first section looks at some of the early projects that the Math Forum was part of beyond their own initial grant. These projects help consolidate the Math Forum as a leading online math education center. The second section I called the digital library era as the Math Forum not only had several NSF digital library grants, but was a key player in the National Science Digital Library (NSDL) project. The third section looks at efforts to advance the notion of mentoring: mentoring teachers, mentoring students, mentoring mentors. Mentoring is a key concept from the beginning of the Math Forum's existence, and these projects advanced mentoring in some important ways. Finally, the last section looks at two large current projects that represent the mature Math Forum and where it might be going next.

Early Projects

Beyond the Math Forum's own original grant, two important projects that it was part of were the Educational Software Components of Tomorrow (ESCOT) project and the Bridging Research and Practice (BRAP) project. Each of these projects was headed by other institutions. Yet the Math Forum was a collaborator with the PIs and the institutions and had an important component of each project.

The ESCOT project was a joint project between SRI International and University of Massachusetts Dartmouth. The PIs were Jeremy Roschelle, Chris Digiano, and Roy Pea from SRI and Jim Kaput from University of Massachusetts Dartmouth. The role of the Math Forum in this project was to put together development teams to produce interactive applets for the Math Forum PoW environment (http://mathforum.org/escot/work shop2000/). The PoWs were called ePoWs and they allowed students to play with applets in order to think practically about a problem. For example, one problem on volume allowed students to pour virtual water between virtual buckets of different sizes in order to think about how to move a certain amount of liquid to a different container as virtual manipulables. Teams consisted of an educational technologist, a software developer, and a middle school teacher. Each member of the team brought a different expertise to the group. The teachers brought experience of pedagogy and teaching math to middle school students. The developers had experience with developing small independent programs that could be inserted into web pages. This would allow math problems to become interactive. Most of the programs were Java applets. The instructional technologists brought expertise in instructional design and the use of technology for learning. In practical terms, often the instructional technologists were mediators between the teachers and the developers who came from really different work worlds.

The ESCOT project allowed the Math Forum to further explore the potential of the internet and technology for learning. The Math Forum had been convinced that different kinds of manipulables gave students different kinds of opportunities to think about math. They developed a problem they called the traffic jam that allowed students to think about movement and algorithms that describe that movement. In face-to-face workshops with middle school students, the Math Forum would do the exercise with actual physical manipulables, such as tiles. Then after discussing with the students patterns they saw they would do the same problem with an applet, virtual manipulable, and after the kids would experiment with that, they would then ask them to think about the patterns. At this point, they would start to lead the kids toward an algebraic expression of the patterns they saw. It was an impressive exercise to watch middle school kids generate algebraic equations using these tools. But the Math Forum staff was also interested in the fact that the different kinds of manipulables, physical or virtual, had a different impact on the groups of kids working together. The virtual manipulables were not just a simulation of the physical ones. And it was not that one set of manipulables was better than the other, but rather that students attended to different aspects of the problem solving with the virtual and physical manipulables and each type contributed to their ability to abstract from the concrete situation. ESCOT influenced Math Forum thinking and paved the way for future digital library projects such as Math Tools and Math Images, discussed below.

The work of the teams was an interesting part of the project in itself. Software developers and teachers lead really different lives. They worked in different kinds of environments with different rhythms. Teachers tended to check email early in the day before school started and then at the end of the school day. Developers were talking through email and chat services all day long. This led to a tension around the work because each group saw the other as inhibiting progress in some ways. Another tension the groups experienced was that between what the developers saw as interesting programming ideas and what the teachers saw as pedagogically appropriate. These were interesting debates and conversations, as I said, often moderated by the instructional technologists. In the end, that made the ePoWs richer and more interesting.

The other early project was the Bridging Research and Practice project. The BRAP project was a collaboration between TERC, Michigan State University, and the Math Forum. The PIs were Ricardo Nemirovsky and David Carraher. The goal of the BRAP project was to bring together the world of math education research and the ways that math education was practiced in the classroom. Understanding that there was a wide gulf between these two spheres, the project asked how each sphere could inform the other but also how technology might help bridge the gap.

As one of the test sites for the project, the Math Forum brought together a group of math teachers from several middle and high schools across the country. Taking the charge from the PIs seriously, the Math Forum empowered the teacher group to take this project in a direction they wanted to go. In an early meeting with the BRAP teacher group the Math Forum staff introduced the teachers to the project but then simply asked them what would they like to do for the project. The teachers were baffled by this request. They had never been empowered to take intellectual charge of a project before, even though they were all strong teachers. It took them some time to wrap their heads around this leadership role, but ultimately they embraced it and decided that they wanted to focus their project on "discourse," which they saw as a critical juncture between practice and research.

The project went on for three years, and this dedicated group of teachers met face to face, had virtual meetings, worked individually on the project, explored the role of mathematical discourse in their own classes, and compared what they were reflecting on in their own practice with the scholarly literature on mathematical discourse. At one important juncture the teachers realized that they needed to videotape their own classes and their practice as teachers. No one on the Math Forum staff led them to that realization; they came to it on their own. While they were nervous about exposing themselves to each other in this way, they realized they needed the videos in order to really objectify what they were doing in class and think about it more deeply. It was a brave and profound moment for them. Ultimately, the teachers helped to produce an incredible video-paper out of the project that had clips from the videos they decided to make (http://mathforum.org/brap/wrap2/).

Digital Library Era

In 2000 the NSF established the National Science Digital Library (NSDL). NSF funded NSDL projects from 2000 to 2011, as a long-running grant program. The idea of NSDL was that there would be a single portal for research, resources, and online communities, in the STEM fields where teachers could go and access these resources (https://nsdl.oercommons.org/nsdl-overview). NSDL was also a dynamic community of researchers who met annually, developed their own digital libraries or DL services, and helped to shape what the NSDL is today.

Math Forum had three main NSDL projects: Math Tools, Leadership Development for Technology Integration, and Math Images. For all three projects, the lead PI at the Math Forum was Gene Klotz. Math Tools was a project where the goal was to create a microcosm of the Math Forum site. Math Tools was a collaboration between the Math Forum and Swarthmore College. Drawing inspiration from the ESCOT project, Math Tools became a library of applets and other small programs that could be embedded in math lessons that would allow students to manipulate virtual objects in order to advance their thinking about some aspect of mathematics. Not only was the digital library a collection of these programs, but also members of the community were invited to rate and review these small programs. Reviews could be critical, but they could also talk about how the program might be used or adapted. Community members were also encouraged to write and share lessons for these programs, and people rated and reviewed the lessons as well. Some of the small programs in Math Tools were singular programs written by individuals for particular purposes. Others were collaborations with groups that had small collections of programs and agreed to share them on the Math Tools site. Math Tools was a major effort to build out the Math Forum site in new directions and in a sense create a subcommunity within the Math Forum online community. Math Tools has been such a success that the discussions on the site have been active years after the funding ended for the project.

Leadership Development for Technology Integration was a Math Forum project set up to create a set of workshops that would allow teachers to take a leadership role in thinking about how to integrate technology into K-12 math courses. The project had a focus on middle grades and high school but spanned the gamut of educational levels. The idea behind the project was for teachers to participate in Math Forum workshops, but then to organize their own local workshops using their personal networks. In this way the work could be disseminated to a larger audience. The project also worked to develop new content resources to be housed on the Math Forum site, so that the digital library would continue to grow and be updated. Inspired by the BRAP project, this project handed over the reins to teachers and supported them in moving into leadership positions both in their own schools and beyond (Regis et al., 2009; Renninger et al., 2011).

Finally, Math Images was designed to create a library of images that were mathematically interesting. The user of the library could explore the images and explore mathematics that was linked to the images. The original location of the site was a wiki space, which would maximize collaboration on the site. In each of these digital library projects the Math Forum was expanding on its original vision in important ways. It was thinking about new ways to build out the resources on the site. The NSDL program gave the Math Forum a new way to think about what it was, an interactive digital library. Further, the NSDL projects gave Math Forum a way to continue to support teacher leadership development. The Math Forum has always looked to strong teachers to both help produce online resources and be leaders in an online community supporting others in their desire to become better mathematical thinkers. Finally, Math Forum took up its place as an important community member in the world of STEM digital libraries. This group was an important community of scholars and practitioners for the Math Forum to work with.

Mentoring Program

Much of the core of the Math Forum is made up of teachers who are strong in math content knowledge and strong in their pedagogic thinking. These leaders contributed resources and supported others in their mathematical development. Many of them mentored students and other teachers in the PoW environment and through Ask Dr. Math. This emphasis on supporting individuals in the development of the problem-solving skills, the ability to talk about mathematics, and ultimately their mathematical thinking led to the development of a mentoring program. Originally, there were two funded research projects, the Online Mentoring Project (OMP) and the Virtual Fieldwork Sequence (VFS), which were designed to support preservice teachers in the process of becoming mathematical thinkers. I was PI of both projects. While this work was focused on pre-service teachers, the insights from this work were generalized to a larger group of practitioners and led the Math Forum to think in new ways about how to support these practitioners.

In these two projects we struggled with the complex task of teaching an attitude toward mathematics and math education. While it could be argued that teacher education for math teachers does not adequately prepare them in both mathematics and pedagogy, the goal of this OMP was not to make up deficits where they might exist. Rather, the goal was to encourage the Math Forum dialectic of engaging in problem solving and talking about that problem solving, with the goal of developing new knowledge or changing the way one thinks about mathematics.

There were many rich discussions that the research team (which included many of the Math Forum staff members) had as a result of these two projects. The insights from those projects not only affected the online materials that were created for pre-service teachers, but they also were used to support in-service teachers as well. At the same time as this work was going on, the Math Forum began to develop a kind of rubric for this work that they called "Noticing & Wondering" (Fetter, 2008; Hogan & Alejandre, 2010). Later I will describe in more detail Noticing & Wondering and its relationship to the literature on teacher professional noticing. But here we want to note that these projects, and other activities during this time, allowed the Math Forum to begin to package its idea of a reflexive practice into a model that might encourage more individuals to embrace it.

VMT and EnCoMPASS

I end this discussion of the history of the Math Forum through the projects that it produced with two current and important projects, the Virtual Math Teams (VMT) project and Emerging Communities for Mathematical Practices and Assessment (EnCoMPASS). VMT is a project where Gerry Stahl was the PI and is now headed by Steve Weimar. VMT is an interesting development at the Math Forum. It is a project designed around a synchronous white board work space and chat function. The idea in VMT was the development of student-driven teams working on mathematics and talking about mathematical ideas together. Motivated by Stahl's (2006) interest in small groups and his passion for group cognition, VMT was the first major project at the Math Forum that focused on synchronous problem solving and synchronous chat, and their potential for students of mathematics. Most of the other work at the Math Forum happens in an asynchronous environment. An important discovery of the VMT project was the idea of the problem scenario. Because the project supported unmoderated teams of students exploring math together, the research team discovered that creating more open-ended scenarios, where students could explore mathematical ideas and then define problems themselves, was much more conducive to the kind of group cognition the research project was seeking to support. This idea of open-ended problem scenarios, as opposed to closed-ended problems that have a particular correct answer (even if there are multiple routes to the correct answer), was valuable to Math Forum thinking about how to encourage people to engage in mathematics as a discourse and way of life.

EnCoMPASS is an ongoing project that takes many of the Math Forum insights over the years in all of the above projects and attempts to leverage those insights to support teacher development. Directed by Jason Silverman (PI), the goal of EnCoMPASS is to take Math Forum insights about developing an online community of teachers and create a new community focused around the ideas of assessing student thinking. As part of the project we developed a software tool that allows teachers to basically analyze student answers to math problems. Teachers are able to highlight sections of work and comment on those sections. They are encouraged to use the Noticing & Wondering (N&W) framework of the Math Forum to notice things about the segment of student work, wonder about what the student was thinking, where they might go with what they have done, and so on. These highlighted chunks of student work can be put into folders that are structured in ways a teacher might want, for example, around math concepts or typical mistakes. The folders can allow a teacher to compare work across students or to share that work with other teachers so that they can discuss questions and issues. By creating this kind of an assessment tool and by building an online community around it, EnCoMPASS seeks to scaffold teachers talking about student work, the mathematics involved, their questions about student thinking, and their ideas about pedagogy. This discourse will further support a culture of mathematical discourse among a community of teachers. It is further thought, along the lines of the community of practice model, that the core members of the online EnCoM-PASS group can mentor and support the development of new teacher members to the community who may be newer teachers of teachers who feel less confident about their mathematical and pedagogical skills.

Conclusion

In Chapter 1, I talked about the Math Forum culture as being made up of three components: the tradition of critical thinking in the elite liberal arts college where they began, the utopian culture of the early internet, and the dynamic personalities of the founding members. Over the course of the many locations and projects they were part of, that culture evolved in some important ways. When the Math Forum was bought by WebCT, it moved off campus, took on new members who were outside the College community, and began to expand its vision of its role. This was an important moment because it was a time when the Forum worked to integrate new outsiders into its community of practice. It also continued to think about who it was as an internet organization and how that should evolve. It has also weathered many financial crises: the pressures when the dot-com bubble burst, the pressure at Drexel to be financially solvent, and then finally the pressure to fit into the RCM budgeting model. This group has had its creative thinking skills stretched to the maximum. Nevertheless, it remains a dynamic and creative group: a community of practice that remains reflexive and balanced.

In their article titled "Teacher Professional Development, Technology, and Communities of Practices: Are We Putting the Cart before the Horse?," Schlager and Fusco (2004) suggest that there has been a lot of talk about communities of practice in education and in technologically mediated educational environments, but ironically there have not been a lot of true communities of practice. They suggest, in fact, that all the talk is there because it has proved so difficult for communities of practice to actually develop. There are many reasons we don't tend to have communities of practice in education, Schlager and Fusco suggest, not the least of which is the fact that there is rarely the trust and the freedom for a community of practice to evolve. Schlager and Fusco are talking specifically about teacher professional development (TPD) in their work, and they point out that TPD is rarely aligned to the actual work that teachers are doing and the needs they have. Rather, the PD is imposed from the outside by some group of experts hired by an administration to make up for perceived weaknesses that exist within a faculty. Further, faculty are often resistant to engaging in TPD because there may be some cultural difference, such as experts supporting inquiry but teachers finding inquiry difficult to use and perhaps even detracting from their perceived sense of classroom control.

While the Math Forum communities are made up of teachers, students, and hobbyists, the core of the online community I would argue is teachers. Teachers are the most persistent group at the Math Forum, and they come to get new lesson ideas, do a workshop for themselves to enhance their professional development, discuss math and pedagogy issues with other teachers, and to have a place for their students to come and do challenge problems to enhance their math learning. As such, TPD and teacher communities have been a large part of the Math Forum community.

Math Forum has been much more successful at fostering communities of practice precisely for some of the reasons Schlager and Fusco say it is hard. For the Math Forum, math is a regular part of everyday life. If we look, we see math all around us. The Math Forum has always been aware that good communication is an important part of anyone doing math. People need to do problems together and to talk about them, and that is the core of its practice. At the core of both stories discussed above, fostering communities is also at the core of everything the Math Forum does. It believes in starting with a person and picking up from what they know and then encouraging them to be part of a social environment that pushes the person not only to problem solve, but also to become better at describing what it is they are doing. It is this *twin focus* that encapsulates Math Forum pedagogy, problem solving, and communicating.[2]

Jean Lave has suggested that the traditional view of learning (for both teachers and students) is a "culture of acquisition" model (Lave, 1997). In that view there is a body of knowledge that has to be transmitted to the learner, and it is really about the acquisition of this knowledge. But in fact, human understanding does not seem to take place in this way. What is a much more accurate view is one that sees knowledge as part of the social forms of consciousness that (as Marx described) are internalized from the forms of practice within the social formation (and the layout of that formation) in which these practices occur. And because of this, the traditional acquisition model fails to explain why more affluent students do better than poor students. Traditional acquisition models fail to explain why teachers with more culture capital are often more able to understand and be creative about teaching material than their colleagues with less cultural capital. The acquisition model does not have this larger social context in mind when thinking about the transmission of knowledge and so cannot see what is going on. For these reasons Lave and others have turned to models of apprenticeship for a better model for learning. The apprenticeship model allows one to view how learning is both social *and* context dependent.

For Wenger (1998), a key piece of the community of practice is a balance between participation and reification. In the apprenticeship model, it is the work of the social group that is important; groups and their work can get out of balance if there is not an equilibrium between the participation and the ways those forms of interaction are "reified" and then become part of the collective memory of the group. Wenger lists helpful examples in his book of how companies develop a form (reification), which grew out of some work process in which they were engaged (participation). He notes how participants of the community of practice then become slavishly committed to the form, even though the context of work changes, as an example of participation and reification getting out of balance. If you take the more classic Marxist definition of reification, Lukács (1972) would argue that in capitalism there is always a tendency for this pairing to get out of balance as capitalism drives to turn social processes into "things."

The way the Math Forum uses technology is an example of a learning organization attempting to keep participation and reification in balance (Wenger, 1998). The wiki spaces and other online spaces, the VMT work space, and the EnCoMPASS assessment tool are spaces where reification occurs. These tools are created precisely because they are mediums that not only objectify social interaction, but also can change quickly and are quite flexible. The objectification can shift as the practices that people are engaged in shift. Further, since these online tools are flexible, they allow new members of the group to see what the more senior members are doing, and these tools can scaffold their participation. *Participation and reification are in a constant dance.* Because the internet allows for the warping of space and time, TPD at the Math Forum can be aligned to teachers' needs, in fact can be led by teachers. Further, it can be kicked off by a two-week summer workshop, but continue online throughout the year.

The central community of practice at the Math Forum reaches out to others and uses the internet to extend that community even further, so much so that one might ask: is the Math Forum an online community of practice or a portal that sponsors many communities of practice? There are teachers and students who use the PoW services all the time, and so one could think of them as peripheral members to the core community. But there are many other outreach groups that are part of the Math Forum and part of that Math Forum culture that, indeed, may be part of other communities. In the following chapters, I provide examples of this rhizomatic structure to the Math Forum community and argue that it is both one community and many communities.

The staff and core participants at the Math Forum are, to my mind, a quintessential community of practice. They have made doing math a craft and they apprentice people in the work of mathematics. The staff is constantly on the lookout, or imagining, new technological tools; they have a complex view of social space, all with an eye to give people more opportunities to engage in problem solving and more opportunities to talk about the work they have done. At the core of this online resource center are people working and communicating together around mathematics.

Notes

[1] An early project that began even before the Geometry Forum, but continued as the Geometry Forum evolved into the Math Forum was a collaboration to help develop and use Geometer's Sketchpad, which is software to allow geometry people to draw and create visual proofs to aid in the solution of geometry problems. This software was sold by Key Curriculum press and illustrated the early creativity of the early Math Forum staff as they worked with the limitations of the internet. It is now part of McGraw Hill Education.

[2] At one point in the Math Forum's work with teachers they were asked to develop a scoring rubric for the work they were doing in the PoW environment. I was against this idea and in fact remember arguing in meetings that what the Math Forum was being asked for was a grading system, which went against what they believed in. The staff listened carefully to my arguments and some of them had similar concerns. They understood better than I that they needed to meet their audience of teachers "where they lived," and so they set out to create a scoring rubric for kids doing problems in the PoW environment. The result was the rubric at: http://mathforum.org/pow/teacher/scoring.html. It follows the Math Forum pattern of focusing on the twin ideals of problem solving and talking about the work that a person has done. The rubric is complex enough that it does do a good job of helping students (and teachers) think more about the math that kids are doing, and at the same time it fulfills the desire for having something like a grade – an assessment of how well the student did with the problem.

4   Possibilities and Their Foreclosure in the Digital Educational Economy

Introduction

If much of the Math Forum practice was an organic and unalienated outgrowth of the idealism of an elite liberal arts institution, the utopian culture of the internet, and the personalities of the early staff members, then that unalienated labor came into direct confrontation with the ideology of the commodification of the digital educational economy. This chapter follows up on that part of the Math Forum history and looks in greater detail at the ironies of a neoliberal view of education in the digital age. The original Math Forum project had two idealistic notions built into it. First, it was to be self-sustaining. In this moment of internet culture, people believed that the internet itself could somehow sustain organizations since communication was inexpensive and digital resources could be reused. In our modern world of massive server farms, this belief could be seen as quite naïve. Second, it was to be a model of other online educational communities. It was thought that creative ways the Math Forum had leveraged new technologies could be taught to other groups.

In the preceding chapter I described how the Math Forum went from being housed at an idealistic small elite liberal arts college to being purchased by WebCT during the dot-com boom to being spun off to another university in the dot-com bust. Now we will look at the broader political and economic context within which all those changes took place. The Math Forum has been both a tremendous resilient internet organization and one of the very successful online educational communities. However, that success has been limited by the ways its history has been shaped by these larger social forces. There are perhaps implications here for the current practices of online education in universities and the coming of the massive open online courses (MOOCs).

Space-Time Transformation in the Information Economy

The history of the Math Forum is tied up in the larger history of the internet and communication technologies as they have developed within the economy of the late twentieth and early twenty-first centuries. There have been multiple volumes dedicated to the rise of the information economy and society (Castells, 2001; Graham & Dutton, 2014; Harvey, 1990; Peters et al., 2009; Schiller, 2000). One overall principle is that of the transformation of space and time over this period of time. In the *Conditions of Postmodernity*, David Harvey (1990) talked about the reorganization of consumer capitalism as it moved from the Fordist/Keynsian model of regulation that begin in the postwar to what he at the time called "flexible accumulation," which was an early version of the contemporary neoliberalism we all know so well today. A critical part of the decentralization of production that Harvey was discussing was the ways new informational technologies were part of producing "space-time compression." At that time, Harvey points out that space-time compression is a long-term part of modernity and that this is an important part of what happens in capitalism; space is rearranged using new technologies to allow for new forms of production and new regimes of capital accumulation. Throughout the development of modern economies, the world has become smaller through new technologies of transportation (e.g., canals, rail, trucking, air) and new technologies of communication (e.g., telegraph, telephone, fax). So the most recent information technologies such as satellite, cellular, and the internet were just the next step in making the world a smaller place. But they had the dramatic impact of beginning to make it possible to make global decisions in an instant, which was quantitatively a huge change.

Harvey set the stage for a new generation of theorists to talk about global flows and the decentered and disjunctive way in which the production of space is being rethought and reorganized (Appadurai, 1990). What we are looking at now is not just the space-time compression of modernity but, in fact, space-time transformation. Space and time can be compressed but can also be expanded and overlapped. Perhaps the warping of space-time might be a more fitting way to think about it.

In an effort to keep this discussion simple, for our purposes we can think of three moments in the contemporary transformation of space-time and their links to conflicts and possibilities in the increasingly more global economy. The first of these phases might be thought of as the restructuring of companies and the deindustrialization of urban North America that new communication technologies made more possible. This first phase occurs a bit before the birth of the Math Forum but set the stage for the world the Math Forum came into. The second phase can be thought of as the internet boom economy of the mid-1990s and then the bust of 1999–2000. This period directly shaped how people thought about the Math Forum and its potential. It also shaped the expectations that would deeply impact the ways that the Math Forum was able to maneuver. Finally, the last period can be thought of as the current phase of increasing globalization, the rise of neoliberal ideology, and the potential of a more entrepreneurial culture.

I begin here by talking about this first phase of space-time transform-ation, and then I will discuss the other phases in the next two sections. Even though it occurred before the Math Forum was formed, this first phase of the development of the information economy and the space-time restructuring that went on during this time is important to help us think about what came next. The post–World War II period saw a period of great economic growth in the United States and one of the longest periods of economic expansion. The United States not only was providing manufactured goods domestically, but it was helping to rebuild the econ-omies of the world. There was great demand for US goods and little competition. But by the end of the 1960s, that expansion reached its peak. Global markets were saturated and foreign competition was increasing. By 1970 there was global stagnation and many industries were flat or decreasing. At the same time, there was double-digit inflation and finan-cial problem due to the OPEC oil embargos and the glut of cash that those embargos created.

In many ways, those economic problems of the 1970s created the world that we now inhabit today. In order to free up markets and provide places for capital to go, there was deregulation of the financial markets, the removal of the gold standard, and the beginning of shifting from manufac-turing to finance at the core of the global system. These transformations have been well documented by many researchers over this period of time (Bluestone & Harrison, 1982; Harvey, 1990; Strange, 1986). Along with the financial deregulation was the deindustrialization of American indus-trial cities. This process began as early as 1969 but was in full swing during the 1970s. By the 1980s sociologists and other social scientists were talking about deindustrialization. The deindustrialization of America changed many things. First, the strength of unions was reduced as companies left the urban North where their workforces were unionized and went first to the American South, but later to other countries. Second, highly paid manufacturing jobs became increasingly scarce, and they were replaced with lower paying service-sector jobs. This was the beginning of the erosion of the working class in the United States and the beginning of the bifurcation of incomes: (1) skilled professionals with university degrees moved into higher paying jobs and (2) individuals from working families who often did not have or could not get these advanced educational degrees moved into the remaining lower paying service jobs (Putnam, 2001).

At the same time, cities began an interesting transformation. In the 1970s and '80s, as cities became less a center for manufacturing and highly paid skilled jobs, they experienced decline. Cities lost population and revenue. By the end of the 1970s the demographic movement was beyond the suburbs and toward the "ruburbs" (not quite rural and not quite suburb), as newspapers were talking about the death of cities. As the industrial economy began to yield to the knowledge economy and the second phase of space-time transformation, cities began to be reorganized as information hubs in a global economy, and so they were restructured and came back as important sites (Sassen, 2001). Initially they began to be important as financial centers in the growing global economy. But then as knowledge and information became so central to the management of the global system, they became important as informational hubs in a much broader sense, especially the largest, most global cities. At the same time, cities sought to attract knowledge workers, and knowledge workers are very different from the old industrial workers. They are looking for different kinds of entertainment, and so cities that were hospitable to the "creative class" – artists, writers, musicians – were more interesting to the professional class (Florida, 2005). Old industrial spaces and worker housing were repurposed for new restaurants, performance spaces, and condominium and living spaces. And so American cities went through reorganization along the lines of the three phases we are discussing.

The Internet Economy

The first phase of the transformation of the urban landscape – deindustrialization, the decline of cities, and then the slow return of cities as information hubs – begins the process of reorganizing urban space for the second phase of space-time transformation: new information economy and new forms of wealth production (Harvey, 1990). By the late 1980s Harvey and others were talking about the ways that companies were changing how they were organized in order to preserve profit. This is Harvey's notion of flexible accumulation. Harvey argued that new communication technologies helped to make flexible accumulation possible. Services such as public relations, which used to be done in-house in most companies, could now be outsourced to independent providers who would contract for those services. These support companies would be paid for the PR work they would do, but would not be employees that would have to be paid during down times when PR services were not critical; the company would not need to support their fringe benefits, such as health care and retirement. In this way, the core of the organization could be quite small and a larger set of temporary employees and service providers could fill in for the work that was needed. This has become such a normal practice in the 1980s and 1990s that by the early 2000s Sassen (2001) states that a good percentage of economic growth in global cities is the work that these support companies provide for global corporations; her point being that it is not economic growth in the traditional sense. Rather, it is the work that traditionally was done within the corporate headquarters but was now being done by these independent service providers.

As we said above, the global economy had stagnated by the early 1970s, and by the end of that decade there were multiple economic contradictions and still much stagnation. It is at that point that the first wave of a conservative political agenda begins to gain general public traction and free market ideology begins to return to a dominant position. The success of the postwar regulation economy begins to yield to increasing demands for deregulation. The welfare state, which was once seen as the ideal system to protect the interests of workers, was now criticized as creating economic stagnation and limiting the futures of workers and companies alike.

At this moment, there were also new information technologies that would allow companies to not only outsource work, as we saw above, but also rationalize activities and automate work that used to require people such as bank tellers, phone bank operators, or supermarket cashiers. There was also increasing realization that there was tremendous surveillance potential of these new communication technologies and that they could be aimed at making workers more productive (Zuboff, 1988). These systems promised to make companies more efficient, less dependent on workers, less dependent on human supervision, more distributed, and more globally competitive. Of course, at the same time, the new world of computers and networking technologies came with a new cost to companies: regular maintenance and upgrading. As a result, company expense shifted from people to machines. Schiller (2000) points out that emphasis on increased information flexibility means that there was a strong pressure for deregulation of much of the telecommunication industry. Telcommunications began to move from a public utility that was in the common interest to a private good that could be capitalized on by affluent companies. It is no accident, and no small irony, that some of the biggest corporations today are the former telecom utilities of a previous era.

The push for telecommunication deregulation dovetailed with the push for financial deregulation in general. As production began to stagnate in the United States and as technologies began to make it possible for production to be distributed, we first had the movement of companies out of the northern and largely unionized cities to the southern states in the 1980s. And then by the 1990s we began to see the pattern of globalization that becomes the dominant pattern in the 2000s where production and consumption are a truly global and mobile phenomenon as companies chase hot consumer markets and locations of inexpensive production.

The 1990s ushered in the boom of the internet economy and the first strong economic growth after the declines of the 1970s and 1980s. It also ushered in a utopian vision of a modern advanced economy beyond the business cycle. With new digital products beginning to appear and companies experiencing the increased flexibility due to new telecommunication technologies, some commentators began to feel that maybe the internet was smoothing out the ups and downs of economic growth. Further, another part of the internet utopia was the idea that the Internet had liberated the creative potential of the individual from social constraint and that the individual and the society could benefit from the unfettered productivity of such individuals. In all of this, there was talk about how the internet may have eliminated the "business cycle" (Weber, 1997). Of course, what we now know is that we were in the midst of the internet boom and the bust was yet to come. But at that moment, it was experienced as something quite different.

During the internet boom, the technology workers and people interested in technology looked like advocates of neoliberal economic policies. They began to suggest that the internet not only had moved us beyond the business cycle but showed us that free and open access was the way to go. While these tech leaders were economically conservative, they also tended to be quite liberal socially. They often saw the new potential of the internet as bringing more voice, great potential for political participation, more freedom and more opportunities for more people. Later in the 1990s and 2000s this would lead to a kind of utopian socialism where individual freedom would create potential for the collective well-being and more people would become interested in open source, the potential of software to deliver goods free to everyone, and so on. And so there would be a kind of shift away from capitalism, and tensions would be created for those who control information-based goods such as software, music, text, or images.

Of course, in many ways this utopian moment of the internet has passed. It passed when the dot-com bubble burst. And like waking up from a dream, we realized that our world had indeed changed, but not in the ways we imagined. It's important to note that while the bust of the internet bubble led to the demise of many internet companies, some companies that have come to represent the information age for us – Amazon, Google, Facebook, Apple – not only survived the dot-com bust but in fact grew during that period of time. WebCT, the company that bought the Math Forum, also survived the dot-com bust, but it was transformed by the change. As we said in Chapter 3, WebCT spun off the Math Forum so that it could concentrate on its core business, the selling of the course management system (CMS) that it had developed. And while WebCT was the company that grew out of the founding of one of the first commercial CMSs, it was bought and merged with Blackboard, a CMS that came a bit later. WebCT itself was a software platform that was created by an academic with academic interests in mind, but it was sold to a larger media company. In 2006 when Blackboard and WebCT merged, one can safely say that the utopian hope of academics was left to create online communities of academics who share interests with each other. And, thus, CMS has become fully part of a commodified vision of higher education. So while WebCT survived the dot-com bust, the company that came later was motivated by a very different set of interests.

Globalization and Neoliberalism

This leads to our third phase of space-time transformation. Along with the globalization of production is the rise of finance capitalism in a much more deregulated financial world. This current moment that developed out of the dot-com bust and into the present is a contradictory moment. We have seen a new economic bubble collapse: the housing bubble and the subprime mortgage crisis. Out of that economic crisis we have learned that much of the economic growth of the last thirty years has been built on financial speculation and real estate speculation, which turned out to be overinflated. Further, we now know that the gulf between the rich and the rest of us has grown significantly and that we are on the road to a new gilded age, if not already there (Piketty, 2014). And at the same time, neoliberalism as an economic ideology and a form of governmentality (Ong, 2006) continues unabated. Many cities continue to develop spaces for the elite knowledge workers and the creative classes, as we have seen the continued growth of the knowledge economy. A number of successful internet companies continue

to become more powerful. We have also seen that in the knowledge economy, universities and other learning organization can play a critical role. We
can start by talking about a number of the success stories after the dot-com
bubble and then move on to a number of other interesting contradictory
lines of force shaping the knowledge economy of the twenty-first century.

Amazon, founded in 1994, was still not turning a profit at the moment
of the dot-com bust, frustrating many of its investors. But following a logic
of internet commerce, what Jeff Bezos and his company were attempting to
do was to grow from an online bookshop to a massive online retailer that
would be the first place people would go for all kinds of shopping. This
building of a massive infrastructure and the continuing reinvestment at
every opportunity was a critical part of that vision. At the end of the 1990s
many commentators understood what Amazon was attempting to do, but
they were still not sure it would work. Just as the dot-com bubble was
difficult to recognize, understanding how to profit in the internet world
was difficult to see too.

Google, MySpace, and Facebook came about right at the end of the
so-called dot-com era. Google, with its impressive search algorithm, was to
bury the competition. But they were to do something else very different.
AOL in an earlier moment cordoned off a small piece of the internet and
said to consumers who were not internet savvy: "Come play here with us.
We will give you the simple tools to log on and you can do everything you
need to do in our little AOL world which is so much easier to navigate than
the larger internet." A short time later, Google was to draw a massive lasso
around a huge chunk of the internet and say to people: "This is not
Googleland, it IS the internet and we give you all of it." Many of us today
live in "Googleland" and think it is the *entire internet*, so they have been
quite successful.

Facebook and MySpace become key web 2.0 platforms to come out of
the dot-com era. While MySpace has faded into a specialized obscurity,
the success of Facebook is interesting. The various valuations of Facebook's net worth have to do with the assumed advertising value of each of
its members. While click-throughs are less common on Facebook than
Google, it continues to have a very high valuation. Other web 2.0 platforms rise and fall on the number of users and the potential advertising
value of those participants. This advertising value, like advertising in
an earlier age, is difficult to calculate, but it becomes the main form of
valuation of platforms.

Ironically in the moment of web 2.0, I would argue that this utopian
hope for the internet and unfettered communication has given way to the

"reification and commodification of the self." This has happened in several different ways. Like the media companies that have come before in the eras of magazines, newspapers, radio, and TV, advertising is the main way that content production is paid for, even if now the distribution is much cheaper. As we have seen in the brief discussion above, participants on websites or mobile social media platforms become objects as they are converted into "dollars per participant," that is, the money they are likely to spend by clicking through an ad, all the way to purchase. If the old media age left us with a dominant discourse that was organized through a logic of persuasion, persuading us to buy this or pay attention to that, the internet has taken the logic of persuasion and made it part of everyone's online interactions. Thus, commodification is part of most everything we do online today. Google and other social media have played a leading role in that transformation.

On another level, we have learned to present ourselves as objects on social media sites such as Twitter, Facebook, or MySpace.[1] While it is now easier to interact on the internet through social media, our interaction is shaped such that we are encouraged toward a particular presentation of self that can include (if we allow it) the endorsement of products and services that show up in our "news feed." Even though many of us no longer know how to build a website, or find a discussion list that is just discussion, without this "framing," we have much more potential for social interaction through these commodified forms. On one level there was certainly a pressure away from DIY and individual control of websites and online spaces. As the technology has become more complex and HTML5 allows pages to be configured whether one is on a PC, tablet, or mobile phone, it also means that more specialized skills are needed to write this code and configure online spaces. Further, as we begin to have a world of millions of servers and to have the power that the online world consumes, there is a drive to figure out how to pay for and profit from this material infrastructure.

This is perhaps similar to the ways that Miller and Horst (2012) talk about the digital dialectic. The internet holds the potential to allow us more unfettered communication with each other. This means that the internet can allow more personal connection and deeper connects as we can stay in touch with friends, coworkers, and loved ones, daily, and even moment by moment. But at the same time the internet increases the number of emails, texts, and other forms of communication we receive. So we are glutted with opportunities to interact with a multitude of others; we can have hundreds and even thousands of "friends" and "followers."

And there is a massive material infrastructure to support this glut. Add to this the tendency for our media industries to monetize these tools so that the workers who keep them going can be supported. The greed of companies squeezes more profit out of these interactions as well. These interactions, which were so personal, are pressured to become more and more abstract.

This dialectic is complex and has left us, in the third phase of the space-time transformation of the information economy, at a contradictory moment. The internet has made it possible for small, craft-based industries to develop. These companies might not have been viable in a previous generation, but now production, distribution, and consumption can be organized with information technologies. That means an individual with an unusual business idea can produce a product or service, sell it across a wide range of people, and make enough money to keep that business going. And so there are lots of small entrepreneurs doing interesting and creative things and helping to transform the twenty-first-century knowledge economy. We now see active and dynamic DIY organizations, musicians and music companies, craft brewing industries, craft food, clothing, coffee, and wine. Many of these companies use the internet and resources in online discussions to acquire the knowledge and structure production, distribution, and consumption of their products and services. This begins to show a glimmer of an advanced economy that gets beyond the logic of scarcity and the image of "homo economicus" (Peters et al., 2009). This is as true in education and educational technology as it is in the larger economy.

At the same time, as Neff (2012) showed during the boom of independent technology companies in the New York City area during the 1999, individuals willingly took on more personal risk and corporations were able to get out from under the financial risk of new ventures by passing it on to employees and entrepreneurs. Further, these same information technology tools have allowed global organizations to become even bigger, to be beyond the reach of nation states and national regulations, creating a growing global elite where a small percentage of very wealthy people control most of the world's information resources and wealth (Piketty, 2014). In the next section I discuss in more detail how the Math Forum has worked to navigate these transformations in the larger economy and society.

Education and the Knowledge Economy

The issues in education parallel those in the larger economy. It's easy to imagine a world where the new information technologies create an educational utopia. Students from kindergarten through the university and beyond have access to the latest scientific and social scientific data. They have access to literature and art, news, and information. These resources can be crafted to meet individual needs and interests. In this world the knowledge economy creates dynamic active learners who can move quickly from novice to expert in any number of areas.

Yet in reality this rarely happens. Information is not knowledge, and the conversion of information into knowing is not a simple task. One problem the internet has created is that it encourages people to conflate a wealth of information with knowing what to do with all that information. To the extent that these resources do meet people's needs and interests, these opportunities tend to fall along socioeconomic lines. This is because students from more affluent backgrounds more often have the cultural capital needed in order to take advantage of information resources (Bourdieu, 1990). Further, the existing structure of our K-12 institutions was built on the industrial model. Classes are designed to produce learners who are of similar skills and abilities. And the increased federal pressures to test students and hold teachers accountable reinforces this factory model of reifying the steps to learning and keeping creativity locked down. As a result, schools have not become dynamic learning sites where information resources have liberated education, except in the most affluent communities.

At the postsecondary level, of course, many new communication technologies were developed here and heavily integrated in research in many fields. There is an odd way in which research has been completely transformed by the internet and new technologies, while teaching has not. Researchers have quick access to articles, data, and each other. Collaboration is now global and work in a number of fields can move quickly. However, integrating university students into this rapidly moving world of university research is much more hit or miss. Some students have the opportunity to begin to experience being part of the knowledge economy and the knowledge society. But many others are excluded from this world. And students at more elite institutions have more opportunities than students at nonelite institutions do. There is a dynamic tension in higher education where entrepreneurial pressures and interests push for a "neoliberalizing" of the university and drawing it more into global capitalism, as it is seen as an institution that provides research for product development and new services. This is true in high tech areas, medicine, pharmaceuticals, engineering, and so on. For those who hold a neoliberal vision of the university, these are the only interesting areas to develop. But at the same time, the creative potential for new ideas and new social arrangements fosters a very different kind of entrepreneurial spirit. This spirit is more tied to a critical vision that might come out of the traditional liberal arts, coupled with the new possibilities that new technologies make possible (Peters et al., 2009). This vision has perhaps more in common with the utopian idealism of the early internet and can also be seen in some creative and successful businesses.

The other big movement in universities has been the effort to reduce the cost of producing educated students. There are many administrations that have hoped that online courses could be the "digital diploma mills" that David Noble (2002) worried about. The ways that universities have imagined the use of the internet for learning is an impoverished view. Rather than thinking about what these new technologies do well and how they can be integrated with other forms of learning, most often they are imagined as an analogue of the face-to-face course. As such, they attempt to provide an equivalent educational "product" that one might get face to face. This is such a common model that now states and the federal government are looking to have universities specify what activities are "classroom" activities and what are "homework," thus solidifying the reification and boxing the online educational experience into a classroom image. This certainly facilitates the "selling" of an equivalent educational product, but it is a far cry from the image we started with – of new communication technologies blurring the boundaries of space and time so that learning and knowledge production can proceed at its own pace, motivated by the interests of the participants.

Of course, there are places for online courses and groups who are doing them well. For audiences that are geographically restricted and need to be transported into a different context, this works well. But in situations where these tools are used well, they are often labor intensive and involve a lot of expensive technologies. They are not the cost savings that some might hope for. MOOCs are one such vision. The vision of the MOOC is to create a high-quality online course that is organized by a leading scholar in the area of the course. In this model, thousands of people worldwide can take such a course and benefit from the digital resources that are placed online. So again, this is an example of the digital dialectic, a course that is removed from the students involved, but one where there are lots of free resources for learners to use, like a library with recorded instruction. Of course, the problem is that learning is a dialogic process. Most people need active interlocutors to engage in the dialogue. Some individuals can take advantage of more static resources, but many cannot. One proposed solution to this problem is a MOOC world with local recitation sessions where learners could engage in dialogue with others. But organizing these local recitations can be a challenge. I began to discuss the issues of learning as a process of meaning making and dialogue in Chapter 2; I follow up those ideas in the next chapter. For now I argue that MOOCs have built on the university's imagination of how the internet can deliver education cheaply to all, but it is a prepackaged notion of education as a reified object that has limited value as currently conceived. So many of the visions of online learning in education have fallen short of what the internet can really provide, ironically because not enough people have looked at the way that the internet has really changed research and knowledge organization and thought about how to apply them to learning in an educational institution.

The Math Forum and the Information Economy

Mark Poster (2001) periodized our media culture into three ages: the age of print media, the age of broadcast media, and the age of the internet. He has a helpful analysis of the kinds of selves that each stage of media articulates and supports. But importantly he thought that the current age of the internet was structured through a new principle of underdetermination. Perhaps similar to the "digital dialectic" of Miller and Horst (2012), underdetermination focused on both the near-infinite reuse potential of internet objects and their near-infinite potential for changing and repurposing (Poster, 2001; Shumar & Madison, 2013). The principle of underdetermination is in a contradictory relationship to the principle of market scarcity that has for so long shaped our economy and the rules we have for interacting in that economy.

One needs only to think about literature or music to appreciate what Poster is talking about. Books, articles, and songs can be distributed freely and unendingly over the internet. Just as important, they can be very cheaply produced by anyone. For example, what drove the cultural market in the past was "studio time"; it was expensive to produce an audio recording. Editing and mockup of a book was expensive to do as well. To distribute these cultural products was costly and one had to rely on existing infrastructure and distribution networks. What made publication houses and record companies valuable was their control of scarce resources. Now in the age of the internet, all these businesses are trying to figure out how to make access scarce by either artificially stopping the flow of information or charging a small tariff for relatively free flow such that individuals will not mind paying.

Our current economic system is built on scarcity. Copyright law was also built on this notion of scarcity. It's not only the way that record company and publishing company executives get paid, but it's the way musicians and writers get paid too. The pressure of the internet to make it easy to distribute all these products and to transform them into new products places pressure on our dominant culture. It not only is a dynamic tension but has produced new tensions. Consumers expect to get music and news, just to name two cultural products, free. This has placed huge pressure on the music industry and has driven many newspapers into bankruptcy.

The Math Forum has found itself caught in this contradiction of underdeterminiation. It worked valiantly to maintain its tradition of a community that shares resources and conversation about math, openly and freely. During the boom internet economy of the 1990s this seemed like it was really possible. One could benefit from the internet's under-determination. The idea that internet resources could be freely reproduced and distributed affected the Math Forum in several ways. First, there was a tremendous sense of hope for the potential of human beings. Math Forum staff members felt that the internet not only would allow people to talk together more and to share resources with each other, but would also allow for a customization of learning and thinking that would benefit all those who participated: students, teachers, and hobbyists. Second, the Math Forum had hope for a new business model where the internet might allow an organization such as theirs to flourish. They were not necessarily looking for profit the way many in the early technology businesses were, but were looking for sustainability. The National Science Foundation (NSF) shared in this second hope. They too began to hope for a new way for projects like the Math Forum to be self-sustaining. The hope was that Math Forum could find a sustainability model beyond the universities that housed them and the granters that funded early projects. Rather, funds from agencies like the NSF would be seed money that would allow groups such as the Math Forum to become autonomous in the larger internet economy. This notion was not fully fleshed out, but talk about the challenges of scaling and sustainability was pervasive. It was believed that if the right recipe were identified, a new way for projects to survive could be found.

It was the boom economy that produced the confidence to develop and support the Math Forum. It was part of the NSF's original investment and was definitely part of WebCT's purchase of the Math Forum. It was also part of the NSF's hope for the National Digital Library project, where there could be lots of online educational communities that could inexpensively produce cutting-edge resources for teachers and students.

This utopian moment, where some believed the business cycle was over, also was a moment when we could imagine ourselves as being beyond an economy of scarcity. In this new internet world the technology not only would bring plenty for everyone, but also would help us overcome our alienation. Information was not just a commodity that could be distributed cheaply, but was a way to connect people to people. From the moment when the text-based internet began, through portals like The Well and the Usenet, to the early web, there grew up a democratic culture with ideas about information sharing and increased potential for people to communicate with each other and all the potential that that brought. Ironically, the hope was the internet could make us more human, even as we become more like cyborgs. The Math Forum culture was very much influenced by this early idealism of the internet. The Math Forum saw the internet as a way to share math resources and make it possible for more people to share ideas and problem solve together. It therefore had an idealistic notion of the communicative potential of the internet. These ideas of free, democratic, and open communication are perhaps similar to Habermas's notions of a communicative ideal (Habermas, 1984). And of course, this utopian moment, while in the past, produced a lot of good and helped us to see a lot of the internet's potential.

Another way in which the underdetermination of the internet affected the Math Forum and its community was its potential for the elaboration and development of the self. Robert Jay Lifton (1993) in the early 1990s talked about the postmodern era and its "protean possibilities." Lifton was talking about the period that others were calling the postmodern, and as such these Protean possibilities for the self were not just a product of the internet. Rather, the sense of the era where an established permanent identity wedded to more stable community and institutional structures was giving way to a more flexible self. This self could be crafted in multiple ways given the more "liquid modernity" that we all were experiencing (Bauman, 2000). Poster (2001) suggests that the self in the internet age is more distributed and is actually part of the network itself. No longer separated and outside, we are process and product in a digital culture. So this new imagined potential for the self was rooted in the historical moment, and many saw the internet as a vehicle for working out interesting self-projects that could contribute to the good of the individual and the good of the community. In our early work with teachers at the Math Forum, identity could be more fluid on the internet (Reninger & Shumar, 2002). Teachers were able to find like-minded individuals who might not be in their school, but over the internet distance it did not matter. They did projects together with these individuals, learned new skills from each other or by taking online training courses together. They acquired ideas to take back to their faculties for local discussions and created online resources to share with each other and a larger world. Since sharing did not cost much, gifts of lesson plans and other resources were freely given. The Internet was a vehicle for a kind of generalized reciprocity (Kollock, 2002).

For the Math Forum, this utopian moment was not just about resources and conversations that could be shared, but also about a new way to create and think about social groupings. The internet opened up new kinds of spaces that were not just hybrid (virtual and face to face) but that had a kind of independent status where who one was (teacher, administrator, student math expert) was less important than how one interacted in the community. And the community was primarily about opportunities to do math together, to talk about math, and to talk about teaching math. The priority of these interactions led to the social groups that cohered around these activities. From the moment we first started doing research with the Math Forum, I have always argued that the Math Forum culture has seen social space as something hybrid and fluid. In the 1990 and early 2000s when so many social theorists were talking about virtual community and meaning something entirely online, the Math Forum was already aware of a more sophisticated view of space and time, much like what Soja (1996) and others might call "third space." The Math Forum staff saw this third space as a space of possibility, where teachers and students could get away from the pressure of their institutions in order to develop their own thinking.

While the Math Forum achieved a lot, their potential as a third space, as a place to take advantage of the underdetermination of the internet, was quite limited. Probably the most important example of that limitation was what happened to the PoWs. From 1993 until 2000 every student who submitted an answer to the PoWs was mentored either by one of the Math Forum staff members or by one of the volunteers who worked regularly mentoring in the PoWs. This made the PoWs a vital service on the internet for math education and could have potentially led to the Math Forum's greater prominence in the math education world. But as we saw in Chapter 3, the Math Forum was pressured to monetize many of their services including and primarily the PoWs. This pressure to commodify resulted in giving up on the utopian potential of the internet and a resignation to the fact that this new way of funding internet organizations never really quite materialized.[2] What is further true, the Math Forum, because it existed in a marginalized space, did not have the support it needed to monetize the PoW services well. And so the result was that the services shrank both in number and in usage. This is not to say that the PoW services are not still one of the most important online math education services – they are. But they could have been much more if a different vision of how to develop them or monetize them were possible.

In summary, the Math Forum found itself caught up in the tension between the neoliberal vision of higher education as an institution at the center of profit production and an institution that could inspire a new vision of entrepreneurship. In the end, because a far-reaching future could not be imagined by the institutions that gave birth to and supported the Math Forum, its growth was circumscribed and the impact it could have on the larger educational world was perhaps limited. But the Math Forum is an amazing organization, and now in their new home at NCTM supporting teachers of mathematics becomes their central goal. It remains to be seen how much further it can go.

Notes

[1] Doubly ironically, many people who did not experience the utopian communication possibilities of the early Usenet now think that Facebook, Twitter, and MySpace are the rails that support a communicative ideal. They fail to see the forms of objectification and reification that these platforms produce even as they are participating in them. Perhaps there is something after all to the Frankfurt notion of the colonization of consciousness through the culture industries.

[2] In a parallel way, some other communication industries have faced the same limitations the Math Forum has in the internet era. Newspapers, magazines, and even record companies are experiencing the same kind of crisis, now that their older funding model is not working as well in the digital era. How to fund the democratic potential of free information in an information society is a contradiction that continues to produce new tensions for all sorts of organizations.

5   Mathematical Conversations and Mathematical Thinking

Introduction

The core of the Math Forum is a forum where people can engage in problem solving and talk about mathematics together. This chapter will explore these ideas in more detail. The Math Forum also became a large digital library with many resources for teachers, students, and families to use. The idea of a resource center became an important part of the Math Forum. It's one of the ways it describes what it is. But at the core of the Math Forum culture is this focus on interactivity and how math is at the center of everyday life.

In the first part of the chapter we will take up Wenger's (1998) notion of the balance between participation and reification. This balance is important to the workings of any group that comes together for some purpose. It's part of what makes a group a successful community of practice. We will move from this idea of the balance between interaction and the capturing of those interactions to a broader discussion of the importance of reification.

The chapter then moves on to a discussion of Math Monday, a practice that was central to the Math Forum staff for many years. The Math Monday story captures some important principles of how the Math Forum sees learning math: it is not just something students do, but also something that most people do. We all use some forms of math in our everyday lives. Finally, we will move to another discussion of the Math Forum dialectic that was introduced in Chapter 2 and a discussion of how that dialectic informed the creation of a rubric for the Problem of the Week (PoW) services.

Reification

A central problem in the social sciences, and one that has become salient in education and the learning sciences as well, is the contradiction of reification. Coming from the work of Lukács and Heidegger, reification names the way that human social interaction tends to be thought of as "things" through a process of objectification (Heidegger, 1927 [1962]; Lukács, 1972). Explicating Marx's notion of commodity fetishism, Lukács talks about the way that the social process of labor, the subjective flow of time, and the uses to which commodities are put are abstractions. The commodity itself appears to be objective. Time is abstracted in the factory, as are labor and value. These social processes melt away, and all we are left with is an object that we fetishize as having the power to move on its own. The Marxist tradition explains well the pitfalls of reification and how consciousness gets trapped within this process of objectification. But reification also does some useful work. It presents objects to consciousness such that the individual can reflect. In fact, human thought would not be possible without reification. I will get back to that point later.

In *Communities of Practice*, Wenger uses the notion of reification to talk about a similar set of operations. For him, reification broadly "congeals experience into objects" (Wenger, 1998: 58). In social organizations, communities, work groups, etc., there is a balance between participation and reification. He gives the example of a business meeting in which there might be a set of notes that are drafted out of that meeting. The notes help individuals hold on to what happened in the meeting. In this example, we can see that the social process is the participation that individuals had in the meeting. The notes are a reification – an object that comes to represent that process back to us. For Wenger, in social groups, there always needs to be a balance around reification and participation. If a group meets on a regular basis but fails to take notes, much from those meetings is lost and even the goal of the meetings might be lost. On the other hand, if we take the notes of the meeting to be the meeting itself, we may end up distorting what we were trying to say or do because we have "fetishized" the notes.

Wenger does not see participation and reification as opposites. He sees them as complementary processes that are part of the way that social groups make meaning. I would agree, but, in general, an overemphasis on reification tends to be the larger problem faced by human groups. This is perhaps because the process of objectification makes it possible for us to reflect, and so we necessarily engage in a distorting process in order to think. It is also true that we live in a commodity-based culture where many of our basic daily concepts – time, labor, money, wages, interest rates, property, and so on. – are viewed as "natural" things and not reifications of human social processes.

Reification is at the core of the way we objectify the world and make ourselves the subject that observes that objective world. This notion is at the core of the way we objectify ourselves. For Anna Sfard (2008), when thinking about how children learn mathematics, a core conundrum is to try and think about how people come to know something new. To know something is to recognize it, and so in theory we should be able to know only the things we already know. New knowledge should be a theoretical impossibility. So how do we come to learn new things? Sfard argues that people proceed from terrain that they are familiar with to new areas through metaphor. In fact, in my last sentence the idea of "terrain" and "areas" are metaphors that help to explain this idea. She draws on ideas from Wittgenstein (1953) as well as Lakoff and Johnson (2003) to talk about the ways that metaphors open up the possibility of new areas. But metaphors, while useful, can also be a problem. The metaphor creates an object that might get in the way of our clearer thinking about what we are trying to understand. Thinking about the expansion of the universe as a balloon is a useful metaphor. But it can also pose a problem for a deeper understanding of the expansion of the universe. The universe, unlike the balloon, is not expanding into anything, distances are just getting further apart. Balloons actually expand into already existing space. The metaphor creates problems as well as helping understand the idea of an expanding universe.

Sfard takes her thinking about metaphor further. Reifications are a kind of primary metaphor. And they happen quickly and at a basic level. As she points out, words such as "cognition" or "number" are metaphors for processes that are more complex. The problem with these basic reifications, as Lukács tried to explain, is that they are the objects that make thinking about these processes possible. We automatically assume that cognition is a real thing and not a metaphor for some more obscure process. Advancing our thought into new areas then is a messy process where we necessarily come up with metaphors, which we then take for the processes we are trying to understand. And then we must deconstruct these concepts and attempt to come up with better ones to move our thought forward. Sfard is clear: reification is the thing that makes thought possible (and I would say, makes possible thinking of ourselves as the thinkers). But it also makes it possible to misrecognize what we are thinking about, because we must think about processes as things. To be able to deal with this might require a particular kind of reflection or critique, where at intervals, we should come back to question the things we think we know. I would argue that this is a kind of "deconstruction" that has become central to Math Forum practice. These ideas will be explored more here and in future chapters.

Reification then is a discursive practice. It is through discourse that we articulate "things" as shorthand for more complex processes than those things come to represent. This shorthand not only makes thinking possible, but it makes it possible to articulate our "ideas" as things in a concise and comprehensible manner. Discourse is always social; there is always a speech community to which some articulation is aimed and within which it is embedded. This is related to Peirce's notion that all thought is dialogic. Peirce (1931) pointed out that all thought has an intended interlocutor. Therefore, even ideas that are developed privately in the mind of the individual have an intended audience. This also means that these thoughts are discursive. In Peirce's terms, they would be articulated through a set of signs that are socially constructed. Another related notion is the idea that learning and knowledge production are processes of intersubjective meaning making (Bruner, 1996; Suthers, 2006). So often what we call thinking is in fact a social process of constructing meaning. This is one reason why Sfard talks about "thinking as communicating," because thought is not a thing, but rather a process of communicating with others in a social group, where the insights and new "knowledge" is produced through the uneven process of using metaphors to explore things we don't currently know.

Communities of Practice

For Wenger and Sfard, reification is always tied up in a larger social process, whether it is a company looking for the balance between participation and reification in its group process or if it is a teacher and a student, or group of students, who are negotiating meaning around mathematical activity. We could refer to these groups, where meaning is made, as communities. In *Situated Learning*, Lave and Wenger (1991) looked at some specific types of communities of practice in order to think about what we could learn about learning from looking at these groups. We could think of these as kinds of "ideal types" where apprenticeship was the model of learning and learning was deeply connected to practice in an organic way. Lave and Wenger picked these ideals not because they were necessarily applicable to schools, but because they contributed some valuable ideas about learning that might be appropriate for schools.

Unfortunately, many educators took the model from *Situated Learning* too literally. People attempted to create ideal communities of practice within schools. What they perhaps failed to realize is that schools are communities of practice and have perhaps smaller communities of practice within them where every classroom might be a community of practice. But as Wenger (1998) shows in his later work, many of these communities are out of balance. The relationship between participation and reification in so many schools is not contributing to collective knowledge production nor to positive outcomes for the individuals and the group. Today, there is a strong drive in education for reification through ideas of assessment and the quantification of quality. We can see with the theoretical thinking above that those reifications become the primary reality rather than the efforts to understand the *processes* through objectification. We come to see the scoring of students and the ranking of schools as the things we think we understand.

Much social learning theory has pushed in a very different direction, focusing on the process. If learning and knowledge are in fact processes where people make meaning together that advances their thinking, then as Sfard and others suggest, we should be focusing on the discourse, the practices, and the ways meanings are being negotiated. What the Math Forum realized is that new technologies can help with these practices because they can be used to facilitate opportunities for people to work and talk together. They produce not only more channels of working and talking together, but talking and working together in qualitatively different kinds of ways.

Math Monday

Several years ago I was flying to Hawaii with a research team from the Math Forum. We were part of a National Science Foundation catalyst project looking into the feasibility of establishing a Science of Learning Center focused on online learning. The various partners of this team were at the University of Wisconsin, University of Colorado, Drexel University, and University of Hawaii. We had taken turns hosting the team and traveled between Philadelphia, Boulder, Madison, and Honolulu. The flight from Newark to Honolulu was about eleven hours and so there was plenty of time to chat and think during the flight. Also several of us, since we were amateur athletes, spent a good portion of the smooth flight standing and keeping our legs moving.

As we were standing in the aisle of the plane, the director of the Math Forum and I began to talk about a math problem. The problem was a classic vector problem where one plane was going one direction at 500 miles per hour with a 40 mph tailwind and another jet was going 500 mph with a

40 mph headwind. It turns out that as you change the numbers, the tailwind does not always make up for what the headwind slows down. As we were talking about how this was counterintuitive, I mentioned that it made sense to me as a runner. I drew the analogy with running uphill and running downhill. I mentioned that one could not make up the time running downhill that is lost by running uphill. While running and airspeed probably have little in common, the conversation continued.

We continued to talk a bit about this the next day at the University of Hawaii. During breaks we talked about these math problems and related problems. I mentioned that I also found it counterintuitive that a 10-minute mile was 6 mph and a 12-minute mile was 5 mph but that an 11-minute mile was not 5.5 mph. This led to more discussion and some calculations. Throughout all of this the director of the Math Forum not only encouraged me to think with him about the math but to do some math as well and to set up the problems in a way that made sense to me.

As an anthropologist I had worked with the Math Forum for a long time at this point. But I am not a person who does math often. I do some problems when I am with the Math Forum, but I would say my knowledge and memory of math is quite limited. But that does not matter to the Math Forum staff, because for them math is something that people just do. It is part of thinking and breathing and being.

After the Hawaii meeting I stopped thinking about runners as a math problem, but the Math Forum did not. They started to take up the problem as part of their Math Monday sessions and began to write problems for use with different groups. In fact, I came to a Math Monday meeting and was amused to see that the running problem was very much part of the discussion that week. They had found a runner's graph from the National Council of the Teachers of Mathematics (NCTM) website, which they incorporated into a set of problems they designed as part of a set of online workshops for in-service teachers. The runner's graph is a little applet that allows the person to start two runners at two different points and then change the pace and see how quickly they run in the same direction and then also run toward each other. The staff then also created another graph to represent the two runners running from different directions toward each other in order to get the problem solver (in this case, the teachers in the workshop) to talk about the information in this graph.

This story is a classic example of the way the Math Forum works. Math Monday was an activity that a core group of staff members participated in for many years.[1] This group included people who work in the Math Forum office as well as telecommuting staff members. As such, it is a social group that interacts within a hybrid (physical and virtual) social space. A group of staff members discusses math problems they are developing for their core PoW services (http://mathforum.org/pow/), as well as for other events, such as workshops. At the time of runner's problem there were four PoW services: Math Fundamentals, Pre-Algebra, Algebra, and Geometry (there used to be more). Math Monday sessions are ones where the staff posts problems to a wiki. Many of these problems come out of their everyday lives: walking to the train station, baking a cake, and so on. The staff members look at these problems on the wiki and have an email exchange over the week about them. Then once a week (often on Monday), they meet to talk about these problems and what each problem needs. The problems tend to be open ended and fairly thought provoking. Their discussions are about math, pedagogy, wording and communication, problem-solving tactics, and more. After each weekly session people go back and continue to work on problems. Usually the original writer of the problem has ownership and brings the problem to publication. When they are published, they will be posted (most likely) to one of the PoW services where students and teachers can work with these problems.

The Math Forum at this time was on the first floor of a small building on Drexel University's campus. The space the Math Forum occupied in some ways resembled the old fishbowl. There was a separate conference room when one entered the front door, and behind that room (which the staff used a lot) was a large open floor. The staff members all worked in cubicles in this largish space. There have almost never been private offices at the Math Forum, and its members' work is always very social. The staff members have become quite skilled at ignoring conversations that are not related to them and listening in when they are connected to the issues. It has served to make them a close working unit. It has also tended to make them quite communal. The group is important, but individuals have less privacy and most work is done in a collective way.

Math Monday takes place in the conference room, as do most meetings. It is a square room with a square table in the middle. There are whiteboards on two walls, bookshelves on the third wall, and the fourth wall is glass block, which is translucent and lets in a lot of light. There is wireless in the room plus many Ethernet cables. Then there is an overhead projector and a flat conference phone system that looks like a flat desktop microphone.

As discussed in the opening story, doing math is at the core of Math Monday. On the staff there are about eight or so active staff members who participate in Math Monday regularly. They do not all write new problems, although most do. Some of the staff members write more than others because they are responsible for one of the four PoW services they are working with and/or they are more experienced members of the group. Math Monday is truly an apprenticeship group. I am probably the member with the least experience and fewest skills doing math (and I am expected to do the math), but there is definitely a continuum from strong and more experienced mathematicians down to newer and less experienced members of the group. Problems are developed by individuals. These may be ideas that were generated in conversation during Math Monday with others. Anything can generate an idea for a math problem, including joking around in sessions that might stimulate an idea and the next thing you know, someone is writing a problem. My contribution of a problem, as I discussed above, was talking about running with staff members on the flight to Hawaii.

All kinds of social situations stimulate math problems. Around the same time that I discovered the Math Forum was working on the running problem during a Math Monday session, there was a new problem the director of the Geometry PoW program wrote about drilling wells, pumping water, and the diameter of the well casing. This problem was in fact a real story right out of the staff member's life, as her cousin had just had a new well drilled (Figure 5.1). She happened to be chatting with the guys who were drilling the well and she learned many details about wells, but also the practical everyday math that well-drillers do as part of their work. A second problem that week had to do with the 2009 Wimbledon tournament and the percentage of first serves and second serves that Roddick and Federer played and won the point. Someone had copied the whole set of data on the two players from their match. A third problem, which was a resurrected from the archive, was about the distance between several staff members' houses and the train station and then taking different routes between the houses (it turns out that three staff members live in the same community). It is not uncommon for the staff to turn to the archive to look for problems that have not been used in a long time. They will pull out an old problem and then redesign it for current purposes.

All of these problems, whether pulled out of the archive or currently written, are put on the PoW wiki. The wiki space is an active workspace. The staff members go to the wiki, work on the problems on their own, and then post notes to the wiki. These notes include everything from wording confusion to their strategies for doing the problem to their thoughts for improvement or for moving the problem in some new direction. The wiki space is really the core of Math Monday activity. Not only do staff members post problems and work on ideas in the space, but also when the

Well, Well, Well [Problem #16027]

I was telling my dad's neighbor Fred about the well my cousin Frank had drilled in his yard. Fred installs well pumps for a living. I said, "It's only producing 3/4 of a gallon a minute, but when it's full, the water is about 400 feet deep."

"That's about 600 gallons of water stored," said Fred. "That should be plenty for his small cabin."

"How'd you know that's 600 gallons?" I asked.

"If it's a modern standard well, it holds 1.5 gallons per foot," answered Fred.

If the cross section of a standard well is circular, what is its diameter?

(Note that there are 231 cubic inches of water in a gallon.)

Figure 5.1 The Well Problem

staff meets for Math Monday meetings, everyone is using their laptops, whether they are in the office conference room or telecommuting. The wiki space is up and part of the work. A common phrase during a Math Monday meeting is "reload": a one-word statement that a Math Forum staff member utters when they have changed the wiki and are asking people to refresh the page so that people can see the changes.

As mentioned above, the group itself is a hybrid group. For the Math Forum, being face to face and being remote is a constant fluid activity. Some staff members are telecommuters and are always phoning into meetings. They may visit the office a few times a year. Other staff members are sometimes off site because they are traveling, doing a workshop elsewhere, and so on. They phone into meetings as well. Math Monday meetings, as well as all Math Forum meetings, have at least a few members phoning into the conference system, Skyping, or Google chatting; sometimes all these modes are working at once. The remote people often have laptops on, and viewing pages is coordinated. The Math Forum is a virtual organization and these technologies are always used creatively to facilitate people's interaction.

The work for Math Monday was reified in a wiki space. The wiki was a place where people could upload problem ideas, work on problems, and raise questions. People worked in the wiki space between meetings, but they also worked in the wiki when they met for their weekly meetings.

Problems in the wiki went from ideas to fully worked problems that were ready for production. In between, there were the traces of mathematical computations that individuals had done showing the way they worked on the problem. Wikis are an interesting choice for this kind of work and an interesting platform in general. By design, they attempt to capture the balance between participation and reification. The wiki workspace not only captures work as it is done, but it can be modified quickly and keeps a history of the changes so that one can see the group's process. Participation in the meeting and a constant refreshing reification were going along together.

Math Monday meetings focused on the math, pedagogy, and language involved in creating nonroutine challenge problems. These discussions moved quickly and staff members often bounced between these areas, but they always understood each other and how these three areas were interrelated. The conversation around math focused on how problems from everyday life, such as the runner problem or the well-drilling problem, could be good for students to think about mathematics. The math had to be interesting, connected to practice, appropriate to the level for which it was being planned (middle school, high school, etc.), and deep. Pedagogically, the staff was always thinking about what level the problem was aimed at, whether it was an appropriate experience for students, and how students could be encouraged to not only get the right answer but to keep thinking about the problem. Language was always critical in the math and pedagogy discussion. Was the language clear enough? Did it encourage people to look at the problem creatively? Was it confusing? These were key questions, among others. The discussion of a problem's math, pedagogical value, and the issues of language were also ways the staff worked to think about how to balance participation and reification.[2]

Math Forum Dialectic

In Chapter 2 we talked about what I called the Math Forum dialectic. The Math Forum's thinking about the interrelationship between the practice of problem solving, discourse about that practice, and the thinking/ knowledge building/meaning making that occurs in the process very much shaped my own theoretical development. In fact, there was a dialectical interplay between the Math Forum's reflection on their own work and my reflections as an ethnographer. We informed each other's theoretical practice. Here I would like to talk more about this Math Forum dialectic as it informed their ideas about the balance between participation and reification: the place of math in everyday life and what learning actually is. The dialectic was part of the practice of Math Monday, and it also informed the kinds of problems the Math Forum came up with through its Math Monday practice.

There are strong cultural norms in mathematics education around learning and getting things right. This is perhaps because in mathematics, it is possible to use incorrect reasoning and come up with correct results. This truth has led much of the structure of math education to be answer-oriented: correct answers are the end results. Many kids who might be "good at math" can get correct answers to problems on tests but not really know what they are doing. Their understanding is dim, but intuitively they have moved toward a correct answer. In the 1980s and 1990s, many in the math education community, pushing for a more social understanding of learning and cognition, began to focus on mathematical practices and process, and not just answers. The NCTM standards in 1989 led a strong charge in this direction and received a lot of criticism from some teachers and math educators for being too progressive and not answer-oriented enough. This began a period of what people called "the math wars" (Mathews, 2005).

The Math Forum, rather like NCTM, always supported the idea that correct answers were important. But correct answers are not everything. It's possible that a student may not know much about the mathematics involved when they get a correct answer. The same debate led to a discussion about procedures versus conceptual knowledge (Hiebert, 1986; Rittle-Johnson & Alibali, 1999). Many of us, including myself, learned procedures for doing algebra, geometry, and calculus; we were considered good math students. But we did not know much conceptually about that math. We did not have many discussions about the math underneath those procedures. What we did discuss, we did not hold on to. Advocates of "mathematical thinking" began to push for more conceptual understanding, which by necessity meant engaging in mathematical practices.

The staff members of the Math Forum, many of them math students in college, automatically took to these more progressive views of math education. They saw both the beauty and truth in mathematics. At the same time, they were sensitive to needing to understand procedures and to show one's work as a road to conceptual understanding. As in so many things, the Math Forum staff attempted to take a balanced approach to the two sides in the "math wars." They recognized that math is hard and requires, as Sfard (2008) would say, going into new areas, but often the language or tools to understand these new areas into which one was going was lacking.

These difficulties can lead to resistance (Cobb et al., 2009), and not only student resistance. Teachers can also become defensive and resistant if they are working on math that they don't understand or have only a limited understanding of. Often, teachers can be driven by a procedural approach to mathematics due to the limits of their own understanding. This can result in missed opportunities to understand what a student is doing, because that student is not following a set of known procedures.

The Math Forum would like to help teachers and students get to the place that they go to when they engage in Math Monday activities. The focus for the Math Forum is on the doing of math. That might involve some procedures, but procedures are always part of the discussion and the thinking about the problem. This is why real world problems, such as drilling a well, are an advantage, because the problem needs to be solved and individuals can talk about different ways to solve the problem and the advantages and disadvantages of different ways to go. In this way, the exploration is discursive and might use reifications and other metaphors to engage in a solution. The dialectical interplay of practices/conversation/ thinking is a natural part of exploring how to solve problems. Of course, with workmen drilling a well or engineers designing an airplane, this is how they always interact with math. But often teachers and students do not have these opportunities, and fear might keep them from exploring this way of working.

So, one of the things the Math Forum has done to address this problem is to design reifications that are meant to lead to process. This is almost a contradictory practice; I would argue it's a kind of deconstruction. In the remainder of this book I will discuss this practice of creating reifications to scaffold process, but in this chapter I will start this conversation with a discussion of the development of the PoW rubric.

The PoW Scoring Rubric

By the early 2000s teachers were asking the Math Forum staff for a scoring rubric for the PoWs. As part of a standard practice in education, many teachers felt that students needed a standardized way to think about how they were being evaluated on their PoW submissions. Teachers also needed to know on what details to focus when producing their answers to problems. Further, the teachers themselves wanted a standardized way to think about the work their students had done for the PoW and what feedback to provide for them, along with any feedback the student might receive from Math Forum staff or volunteers.

As a result of these requests, the Math Forum staff convened a series of meetings to discuss how to develop a scoring rubric and who would do the work in this development process. There were several meetings held to look at different iterations of the rubric as it began to take shape. By this point, I was a regular member of the Math Forum culture. While not a staff member nor an expert in math education, I was beginning to be viewed as a valuable member of the community and an individual who shared the Math Forum's view of a process-based learning style and the things that technology could do for that practice.

I was dead set against the rubric. While we were not in the practice of using the term "reification" in our everyday conversations at that point, I essentially argued that a rubric was the opposite of what the Math Forum stood for. At the time I felt that the Math Forum's process orientation and focus on problem solving, talking, and thinking about math meant that it did not want to produce a rubric. To my mind a rubric was about "right answers" and focused on the product, not the process of doing math. Because I had become a "good friend" of the Math Forum at that point, I did not yield to my normal ethnographic tendency of holding my tongue and seeing what happened. Rather, I jumped into the fray of debate. I was not the only one who had concerns. I added to the voice of what I would call a minority opinion. And others certainly were sympathetic to the concerns, but they continued to work on the rubric and think about how to produce one that would work for the PoWs.

At the time, I saw this as the Math Forum yielding to the pressure from schools and teachers for an answer-focused way to work with the PoWs. No doubt, this is what motivated a lot of teachers in their request for a rubric. I saw the Math Forum's effort to score students as Novice, Apprentice, Practitioner, or Expert (Figure 5.2) as a thinly veiled effort to disguise the fact that this was a 1–4 Likert scale. Basically, I gave up pushing on the Math Forum because I respected that this was a compromise they needed to make. While I was aware that work with the rubric continued and that they refined it with content for the different subject areas, I stopped thinking deeply about the rubric.

If one looks closely at the rubric, and if we think about how the rubric is used, we can see that it is not so much about scoring and is more about scaffolding a way to think about how to look at student work. Only one part of the rubric is about correct answers: the section in Problem Solving on accuracy. Even accuracy is much broader than a correct answer. Further, while problem solving is listed as first on the rubric, and it could be interpreted that problem solving is the most important, half of the rubric

How Submissions Are Scored

We look for good problem solving *and* strong mathematical communication when reading submissions to our Problems of the Week. Your solution should include enough information to help another student understand the steps that you took and the decisions that you made in solving the problem.

Submissions are scored using the following categories:

Problem Solving

- Interpretation: interpret the problem correctly and attempt to solve all of the parts.
- Strategy: pick a good strategy and apply it well–achieve success through skill instead of luck.
- Accuracy: get the calculations and details correct, including writing correct statements and equations.

Communication

- Completeness: explain all the steps taken to solve the problem.
- Clarity: explain the steps in such a way that a fellow student would understand, and make an effort to check formatting, vocabulary, and spelling.
- Reflection: check the answer, reflect on its reasonableness, summarize the process, and connect it to prior knowledge and experience.

Submissions are scored using four levels of performance.

■ **Novice:** "Just starting out"

□ **Apprentice:** "On the right track, but not quite there"

■ **Practitioner:** "Got it"

■ **Expert** : "Wow! Above expectations in some way"

Figure 5.2 The Problem of the Week Scoring Rubric

is dedicated to communication. Overall, the rubric gives the user a way to think about doing mathematics in general. There is problem solving and there are different aspects to the work that goes into problem solving. And there is communication about the mathematical work that one is attempting to do and there are different aspects of that communication. In this way, a person can be doing well with one part of the work on math problems, but needs to develop their skills in other areas. Everyone is in process, because it would be odd to be an "expert" in everything. Even if one was an expert in everything, the narrative response to the student from the mentor or teacher would encourage further reflection. It turned out I did not see what the Math Forum could do with the idea of a rubric, and I was too "reified" in my own thinking about rubrics.

In the next problem, we can look a bit more at how the scoring rubric is used. Figure 5.3 again presents a problem from everyday life. A middle school math club contributed the problem.

While this problem is not completely open ended, we can see there are open-ended elements. The fact that the student is asked, "How much of this can you figure out?" implies that we cannot know everything that is being asked. Further, the "Extra" portion gives the student the opportunity to reframe the problem and find out other things, for example, if there were

Math Club Mystery [Problem #4036]
Members of the Math Club at Morganson Middle School sent us this problem:
The math club took a field trip to see the movie *An Inconvenient Truth*. A total of 28 people went on the trip, including students, teachers, and parent chaperones. There were more parents than teachers.
Movie tickets cost $7, but the students got a discount and only had to pay $3 each. The group paid a total of $108 to get everyone in. Determine how many students, how many teachers, and how many parents went on the trip.
How much can you figure out and be certain of about how many students, teachers, and parents went on the trip?

Extra: State a mathematical relationship between two of the types of people (students, teachers, parents) that would allow you to be certain of all three answers if it had been part of the original problem.

Figure 5.3 Math Club Mystery Problem

other parameters given. Over time, the Math Forum moved from closed- to more open-ended problems. But we will discuss that later.

Figure 5.4 shows the work of one student who has decided to do the problem. The student first gives a specific answer, and then explains her answer in some detail. This follows the Math Forum general culture in which one has to talk about the mathematical work that one has done. We can see here that if we were looking at this from a more traditional perspective, we might see this as a pretty complete answer to the problem. But, of course, from the perspective of explaining what one knows about the mathematics involved, there is something missing from the explanation. Work is shown, but we do not know what the student knows from what we see.

Figures 5.5 shows the reply of a teacher who is using the problem with his class: he is mentoring a particular student who has answered the problem. What you don't see is that the teacher has had a detailed inter-action with a Math Forum staff person. The Math Forum staff and teacher have talked about how to respond to this student, and when they were both satisfied with the conversation, the finished response shown in the figure was sent to the student.

First, if we look at the scoring rubric, we can see that the decision was to rank the student fairly high on all the problem-solving categories. While the mentor is still not entirely sure what the student is thinking, he can see that there is a plan and the work has been laid out well. On communication, more variation exists. The student is rated as an apprentice for complete-ness because she has not explained much of what she has done. While what

Answer:
You know that there are exactly 22 students. But you cannot be sure of the exact number of teachers and parents.

Explanation:
I made 2 equations. One for the amount of people going, and one for the amount of tickets. Then i found what the value of x was, and substituted it back into the other equation.

$x + y + z = 28$ x = parents $7x + 7y + 3z = 108$
                         y = teachers
                         z = students

$x + y + z = 28$
$7x + 7y + 3z = 108$ $x + y + z = 28$
$x = 28 - y - z$

$7(28 - y - z) + 7y + 3z = 108$
$196 - 7y - 7z + 7y + 3z = 108$
$-7z + 3z = -88$
$-4z = -88$
$z = 22$

Figure 5.4 Student Solution to Math Club Mystery Problem

she does say is clear, she is rated the lowest on reflection since there is little effort to reflect on what has been done. As a result, we can see this work is in process: some of it has been done well, yet other things need to be improved.

If we then look to the teacher's narrative, he does several novel things. First, he points out that he arrived at the same answer, twenty-two students and six adults. This is an interesting modeling strategy, as the student said there were twenty-two students, but she did not say there were six adults. The implication is there, because $28 - 22 = 6$. But she did not state it because she could not determine which adults were teachers and which were parents. By doing this, the teacher shows that while the student is right in stating the number of adults, she cannot say how many were parents and how many were teachers.

The teacher then goes on to talk about how he solved the problem and that if the student would tell him more, they could figure out if they were thinking about the problem in the same way. *This is double reflection*: a creative way to say that the student needed to explain her work better. Not only does the teacher ask for explanation, but his response shows that mathematics is a conversation. If she tells him more, then they can discuss further what they did and come to new understandings that this problem makes possible.

Reply by teacher's Name ▼ on Date and Time of Post

Scoring grid will be visible to students.

Score

	novice	apprentice	practitioner	expert
Problem Solving				
Interpretation			X	
Strategy			X	
Accuracy			X	
Communication				
Completeness		X		
Clarity				X
Reflection	X			

Message:
Hi Student's Name,
We both got the same answer, 22 students and 6 adults. I like how you pointed out we know the number of adults, but cannot be sure how many were teachers and how many were parents. I was wondering if you checked your work and if so how? I was able to follow exactly what you did with your explanation.

I solved the problem a little differently though. If you could explain to me where your equations came from and what thoughts were behind them, I would be able to tell better if we had the same line of thinking and we just came up with different equations.

I used only two variables. I made a variable for adults instead of one of teachers and one for parents. Does this make sense to you? An excellent way to check a problem is to solve it two different ways and get the same answer. You seam to understand very well what is going on in this problem and I think it would be a fun challenge for you to solve the problem using only two variables. I would love for your input on which way you think is easier and which way would be better to teach to a student who is struggling with solving this problem.

I look forward to your revisions!
Thanks,
Teacher's Name

Note for Approver:
Aprover's Name,
That is definitely useful. The tone in your response makes the student feel like she is on level with you and is helping you out by doing more. I know if I were a student and received both of our response, I would be much more likely to revise if I received one like you modeled. I tried to do it myself but it was tough to not just copy what you wrote since it was for this problem.

Figure 5.5 Reply from Teacher Mentoring a Student Doing the Math Club Mystery Problem

Finally, the teacher leaves a note for the Math Forum staff member who is mentoring him. He reflects on what he thinks might have motivated him to work more on a math problem. He realizes that the tone the Math Forum mentor has encouraged him to take levels out the status differences between teacher and student. That is an important part of making a free-flowing conversation possible. He then admits his own dependency on using his mentor's words to help him move his mentoring forward. Everyone is scaffolding this movement forward by words that are not their own.

While this is a good example of how the scoring rubric can work well and be integrated with narrative feedback to a student who has done the problem, it should be pointed out that it does not always work. Sometimes there are students who provide a minimal answer to the problem they are working on, and the feedback they receive from a teacher mentor is sometimes not very detailed. There is no magic bullet for making mathematical interactions richer. But it is clear that the Math Forum always has a strategy that is linked to its core ideas about the relationship between problem solving, conversation, and thinking.

Conclusion

Central to the notion of reification is the Hegelian idea of objectification. The interesting point of objectification is that creating objects of thought is, in fact, what makes thought concrete and, in a sense, possible. It's what allows us to say we have knowledge and that we have learned things. We make objects that allow us to reflect on ourselves through them (Miller, 1987; Sfard, 2008). In a real way, self and object are produced through this process. This is Sfard's point: this process of objectification is necessary to learn new things and for knowledge to be developed. Yet we are always at risk of being trapped within the objects we create for this purpose. As an educational psychologist, Sfard is usually thinking about the conceptual traps that K-12 students get caught in and how it might be possible to support better math education. This is the case for many of the math educators coming from a social learning perspective.

The Math Forum shares this intellectual heritage and the Hegelian view of the world. But at the same time it is composed of practitioners who are most often working on these ideas in practice, and so the philosophical notions are not unconscious, but beneath the surface. I never heard anyone give a philosophical justification for the Math Forum PoW scoring rubric. Yet the staff was very attentive to what the rubric might do to students as they are in the process of doing the problems and what it might do to teachers who are reflecting on the students' work.

The scoring rubric makes it easy for students and teachers to reflect on some important pieces of the problem-solving process. It scaffolds their movement from a product orientation to a process orientation. It's a reification that seeks to move people past the object and toward a more dynamic processual view of mathematics. It seeks to erase itself as it supports individuals in their mathematical work. It encourages a deconstruction. Instead of focusing just on specific work and right answers, the rubric nudges teachers and students toward mathematical practice that is self-conscious and leads to reflective discourse on that practice and new ways of understanding math.

Notes

[1] The staff stopped doing Math Monday for a while as things began to change with their situation at Drexel. Work on problems became more informally organized. It remains to be seen if they start up Math Monday again at their new home at NCTM.

[2] As one can see looking at the Well, Well, Well problem in Figure 5.1, there is a specific answer that the problem is seeking. So PoW problems are answer oriented. But to get full credit, students must show their work in detail and communicate well about the problem. Further, mentors always encouraged students to keep thinking about the problem. In these ways the PoW problems are very process oriented. In later chapters, we will discuss how the Math Forum developed the problem scenario as a way to further open up problems. The scenario creates a context where the student must create problems and answer them, giving the participant more control over the knowledge development process.

6 Mentoring Students and Faculty with Digital Technology

Introduction

It's certainly well understood that learning mathematics is difficult for many students. There is significant complexity to deal with. First, there are many new concepts that have to be explored somehow, even though students have no experience with those concepts. Students must use metaphors as a way to move their thought forward while potentially overinterpreting those metaphors they use to develop their knowledge, as we discussed in the last chapter (Sfard, 2008). Further, mathematical knowledge is hierarchical in that concepts build on prior concepts. And if one's foundation is weak, then one's understanding of future concepts is likely to falter as well. Finally, as we know, in the broader educational culture there are traditional ways of teaching and learning mathematics that are procedural. The culture sees math as an individual and cognitive activity, not a social activity that involves discussion and making decisions (Boaler & Greeno, 2000).

Teachers themselves were once students of math. And many of them as students went through this traditional approach to learning mathematics. Some of them excelled and are now confident in their mathematical ability. Not surprisingly, many of these teachers are upper-level high school teachers. But even these teachers, with mathematical confidence, may have a procedural approach to pedagogy and not a lot of ideas to support students who are struggling with math.

Many other teachers are less confident in their mathematical ability. As students they were either weaker in math or at best average. As Boaler & Greeno (2000: 171) state, many students surrendered themselves to a procedural approach and did not have much "agency" in the work they did in mathematics. These teachers can get through math lessons, but they often do not have a lot of advice to give students and they often miss opportunities for teaching because they are not thinking mathematically but rather following a routine. Finally, they know their own deficits and so avoid thinking about them in order to maintain face in their schools and classrooms. Many of these teachers are teaching the foundational concepts that will be needed for later work when the students encounter the teachers who are more confident in math. Then many of these students will need to avoid the threats to their own identity when they have trouble following the work (Cobb et al., 2009).

As more researchers and practitioners began to think about the social context of learning in the 1980s and 1990s, more began to advocate a processual approach to teaching and learning mathematics. But a processual approach is difficult to enact because many teachers, even teachers who are strong in mathematics, are uncomfortable with the approach. Many students don't know where to start in a discourse of thinking about math. And teachers and schools have objectives for teaching that they must meet. The Math Forum, conceiving of itself as a support for math education, had the luxury of thinking about how to support students and teachers to move to a more process-oriented approach – to talking, thinking about, and doing math. As a support community, it did not have the same curricular pressures that schools face. And as composed of people who are passionate about math and how math is rooted in everyday life, the Math Forum could approach mathematics in an organic way.

The notion of scaffolding has grown out of Vygotskian psychology, the work of Jerome Bruner, and others thinkers from a social learning perspective. Scaffolding became an important way to think about how to help people move away from learning procedures and toward a way of thinking about mathematics (Bruner, 1966; Pea, 2004). For Bruner, the scaffold was a way to access the notion of the zone or proximal development. Kids could learn things more deeply and earlier by having a support to allow them to work on ideas that they did not know. In his *Theory of Instruction*, Bruner showed how introducing mathematical concepts first with concrete manipulables like wooden blocks, and then by encouraging kids to develop a language to express the patterns they saw in the blocks, could move those students toward abstract algebraic expressions at an early age. The scaffold then, itself a metaphor, is yet another way to refer to the development of knowledge and learning using metaphors as supports for moving into unknown areas of understanding. And while the early research was concerned with work with schoolchildren, the scaffold could be a support for learning new ideas at any age and at any level.

The theory of scaffolding led to focus on individuals – mentors – who could help provide those scaffolds and manage how a scaffold might most effectively be used. Pea (2004) suggests that a theory of scaffolding would be intimately tied with one's theory of instruction, that is, how one thinks about the learning process, the kinds of support for learning that could be offered, and so on. He importantly suggests that part of the work of a scaffold is to know when the building is built and it is time to take the scaffold away, and perhaps how to do that. Scaffolds and scaffolding are metaphors for the process of making sense of things, a process of meaning making. Finally, Pea suggests that technological tools could very much be part of this process, but in our theory we would need to think about what work software could do, and what work would require more direct human interaction.

Mentoring at the Core of the Math Forum

I have already discussed in earlier chapters the Math Forum's dialectic of engaging in mathematical practices that are part of everyday life, mathematical conversations that make sense of practice, and then mathematical thinking. The dialectic is a process of continual reflection and adjustment. Because of this orientation toward mathematics, the Math Forum staff has been very concerned with the ideas of scaffolding. In fact, we have seen that the Math Forum's own problem-solving rubric is a scaffold for both teaching and learning. At the core of everything the Math Forum does is mentoring. And it is an organization that produces layers of mentoring, mentors who mentor future mentors of students, for example. Further, this concern with mentoring and the layers of mentors has led it to be centrally concerned with the question that Pea (2004) raised: how can software help in this process of mentoring and layering mentors? The early workshops that the Math Forum ran were meant to be sessions where they mentored teachers to become key members of the Math Forum community and contributors to a larger world of math educators. The hope was that these early adopters would then mentor other teachers, as well as students and parents, in math, and the Math Forum would be a hub to advance thinking about math. In the same ways, the founding two services, Ask Dr. Math and the Problem of the Week (PoW), were online mentoring services. While we have discussed the two services before, let's take another look at them from the perspective of mentoring.

In the early days of the worldwide web, there were a number of question and answer services. Q&A was one of the first things that the internet did well. As many in popular culture have jokingly pointed out, we no longer need to debate facts with our friends and colleagues; we can simply look up the answers on Google. And certainly the combination of Google and the use of Wikipedia as an informational resource has come to define a lot of online Q&A. Along with a lot of organizations, Ask Dr. Math was the Math Forum's key Q&A service. And in the days when Q&A was so important, Math Forum was best known for Ask Dr. Math. Ask Dr. Math won a number of awards and was written up in a number of popular periodicals. Ask Dr. Math was so well known that sometimes the staff complained that people had never heard of the Math Forum but they knew well the Ask Dr. Math service. While many other Q&A services have disappeared, Ask Dr. Math is still an important service in the world of math education.

Central to Ask Dr. Math was the scaling question. The Math Forum had many questions and needed a lot of math doctors to answer those questions. Ask Dr. Math was built around a process of training mentors. Mentors might be teachers, university faculty – there were even a few precocious students who became mentors. These mentors were themselves mentored by other math doctors. The Math Forum used the metaphor of a tenuring process. Mentors were scaffolded to provide good answers to students. Eventually, when a senior doctor felt that the junior math doctor was ready, that person was tenured, and they were able to send replies to students on their own. The scaffolding process focused on a number of key areas that one sees in the literature on scaffolding (Pea, 2004).

In the tenuring process of math doctors, there were several things that the senior doctor was looking for in the doctor-in-training. First, there is a focus on language and politely engaging the questioner in the issues of the question. Then answers are provided that are detailed and clear, with the math work shown. In the example in Figure 6.1, we can see that the student has a fairly complex question about decimals and fractions. We can see that the math doctor, one of the senior doctors, has provided a clear explanation about the principle behind converting decimals to fractions and the reverse, but also has worked through the two examples that the student has asked about in detail. The math doctor shows great interest in this work and presents it in a clear and interesting manner. The math doctor leaves some work for the student to do, having provided an excellent model to follow. The doctor then asks the student if this makes sense. The student did in fact write back, saying that the answer was clear and helped the student understand the concept well.

We can see that there is some interesting potential here when using the internet for this kind of scaffolded support. First, there is no time pressure

33 1/3 Percent as a Fraction

Date: 08/03/2003 at 12:50:41
From: Nerd
Subject: Percents

I need help turning percents like 33 1/3%, 2.5%, and other percents like that into fractions. Can you help me?

Date: 08/03/2003 at 15:47:31
From: Doctor
Subject: Re: Percents

Hi Nerd,

The thing to remember about a percentage is that it's just a fraction with a denominator of 100. That's all it is.

So 22% is exactly the same as 22/100. This means that 33% is the same as 33/100, and 33 1/3% is the same as

```
33 1/3
------
 100
```

Now, this is a pretty ugly fraction. It would be nicer-looking if we multiplied the numerator and denominator by 3, which changes the appearance without changing the value:

```
33 1/3   3   100
------ * - = ---
 100     3   300
```

And this can be simplified to 1/3.

Now, what about 2.5%? This is the same as

```
2.5
---
100
```

This would be nicer looking if we got rid of the decimal places, which we can do by multiplying the numerator and denominator by powers of 10:

```
2.5  10   25
```

Figure 6.1 Example of an Ask Dr. Math Reply to a Math Question

for the student or the mentor. Each can take the time to become clear about what the question is asking and what the answer should look like when it is clearly laid out and explained in detail. There can be a dual focus on problem solving and clear communication. The student can be encouraged to look at more detailed mathematical work because the model is before them and they can return to it. It does not disappear; they can go back to it and rethink the interaction. Further, any student with the same question can reuse this mentored interaction. Persistence here plays an important role in the scaffolding process. Of course, a student could return with follow-up questions, and if they made a good contribution to thinking about the question, they too would be added to the archive of the question.

Throughout the history of Ask Dr. Math there was a layering of scaffolding and providing good models that people could follow. As I have said, doctors were mentored to become independent doctors who could engage students with good responses to questions. Students were mentored by the doctors to support their thinking about the questions they had asked and the mathematics involved in the answers to their questions. Finally, good questions were chosen to be part of the archive, and the best answers to those questions were provided in the archive. Then users of Ask Dr. Math were scaffolded by the interface to search for answers to their questions or browse through questions and answers before asking a new question. All of this is motivated by the way that the Math Forum thinks about math as a process of doing and communicating and the Internet as a tool that facilitates the sharing of work and ideas.

The Problem of the Week (PoW) services were also organized around mentoring and involved many volunteer mentors working with the solutions that students provided to the weekly nonroutine challenge problems. In the early days, most of the mentors for the PoW were volunteers and staff members. But as the PoW services evolved, teachers in a classroom, who used the service, were mentored by the Math Forum staff to be the mentors for their own students who were using the problems. This training of mentors also followed the same model as Ask Dr. Math. While mentors were not "tenured" as such, they eventually got privileges to "direct send" to the student. This meant that they no longer required a staff member or some senior mentor to look over their response to a student and approve it before they could send out that response. As we saw in the Figure 5.5 in Chapter 5, the teacher, who was mentoring his or her own students in the class, had an "approver" who was helping with the replies. The approver worked to scaffold the work of the mentor. In Chapter 5 we looked at this

Memo:

Hi Teachers Name,

I agree with your assessment of Student Name's work and like the topics you've chosen to focus on with her -- reflection and explanation.

Since you are asking her to explain how her formulas were set up, I wonder if her Completeness score should be Apprentice?

I think it's ok to just tell her you want to see how she came up with the formulas. Especially if you can give a short sentence about why it's good to show how you came up with them.

With this portion, "How do youknow your answer is correct?You said in your answer that you can't be sure of the exact number of teachers and parents.With this in mind, see if you can solve it another way by revising your formula for the money from the tickets sold." imagining myself as Student's Name, I wasn't sure how to address that... Didyou mean that there was an error in her formula? Or that if she used a different strategy she could find the exact number of students and teachers? I didn't see an error in her algebra or another algebraictechnique she could use. It seemed to me like she had done all the solving she could, and needs to start thinking about the other constraints in the problem and how she could narrow down her answer.

Once she has her answer narrowed down, what specifically would you like her to reflect on? What did you think were the interesting challenges or math ideas in the problem?

Thanks!
Math Forum Staff Person's Name

Figure 6.2 Approver's Conversation with Mentor for the Math Club Mystery Problem

example in order to talk about the role of the scoring rubric and how mathematical practices and conversation were supported by the rubric. But as we are talking about mentoring here, let's return to the same example and look at a piece of the story we did not look at last time. Let's look at the interaction between the mentor and the approver in Figure 6.2.

We can see in Figure 6.2 that the approver, who is a Math Forum staff member, is giving the teacher some advice about what the teacher plans to provide to the student. First, there is some discussion about how the teacher scored the student on the PoW scoring sheet. This is a simple way to start the conversation before getting into more complex topics. The Math Forum staff person then goes on to harder material, showing how the teacher's wording could be ambiguous and imply that the student is wrong. The staff member, as a mentor, also points out that there is no scaffold in the response and it does not give the student a way of moving forward. The Math Forum staff person does this by indicating that if they had received this feedback they might not know how to move forward. This is an interesting choice of how to share this with the teacher. The staff member mentor, by using him- or herself as an example, encourages the teacher to take up a reflexive position. Indirectly the question is asked, "How would you respond to this feedback and so how is the student likely to respond?"

The Math Forum staffer then goes on to ask the teacher what he or she would like the student to reflect on and what the challenges in the problem were.

The Math Forum staff member and the teacher had a couple more interactions behind the scenes around responding to this student. The teacher indicated that he liked the ideas that the Math Forum staff member had suggested but was not exactly sure how to reply. The staff member then said that they wanted to try something and wrote a sample reply to the student, essentially saying, "I am not sure exactly what you were thinking. If I had seen a bit more of your work I might have a better idea. I was thinking this, were you thinking the same thing as me?" This kind of reflection does not critique the student but rather asked them to reflect on their internal process. The staff member told the teacher he could use the reply to model his own replies, but he did not have to use it at all. The teacher thanked the staff person, and used the same basic structure, replying to the student with some of his own worlds. The teacher then reflected on the value of having this model for thinking about how to reply to students.

This was a very dense and complex few turns of mentoring. The focus was on the Math Forum principles of the doing of the mathematics and the communication about that work. Following principles of good mentoring, the staff person encouraged the teacher to take the student's position and reflect on how to respond to the feedback, what they might need to move forward with their work, and target some specific areas to work toward. At the same time, the staff person attempted to model good feedback by thinking about how the teacher must be receiving this feedback and what the teacher needed in order to be moving forward. The staff person here is modeling a process of double reflection, reflecting on your own process as you reflect on what the student is doing. The Math Forum staff is practiced in the art of providing good feedback. They have written their own guidelines to mentoring that connect well to basic ideas in sociocultural learning theory.

Technological Affordances of Mentoring Pre-Service Teachers

As the Math Forum developed, it did many workshops with in-service in both the summer workshops and workshops associated with NSF projects. They also went to school districts and worked with teachers in shorter workshop sessions. Over time, the Math Forum began to think that its resources were not only useful to teachers, but could be good training for pre-service teachers as well. As the literature in math education began to support the importance of a more process-oriented approach to working with math and the need to focus on the evidence of what students were doing, the Math Forum staff realized that the resources at the Math Forum could be helpful to those new pre-service teachers, especially the ones who were lacking confidence in their mathematical ability.

Aware that they were not able to make up for deficits in pre-service teachers' math education, the Math Forum developed a project that was focused on developing an attitude toward mathematics that would allow pre-service teachers to explore basic mathematics with their students. This project was called the Online Mentoring Project (OMP). The goal of OMP was to help pre-service teachers think about helping students in problem solving in what the Math Forum staff called a "pre-field experience." The idea of the pre-field experience is an interesting one. The project imagined a virtual space where university students could get trained to think about how to interact with younger pupils of mathematics. That training would happen online, and then the pre-service teachers would be able to mentor the pupils in the PoW environment. Without the regular distractions of a classroom – for example, the needs of other students, the need to maintain order – the pre-service teacher could focus on what work the pupil had done, how to interpret that work, and then what support to offer that student. The project originally imagined all of this taking place within a web portal that then could be expanded to allow pre-service teachers from different universities to share their experiences with each other. But due to budgetary limitations, the vision was scaled back to producing a curriculum in WebCT that then linked to the PoW back office.

The OMP project created an Online Mentoring Guide (OMG) in WebCT. The OMG had three elements: problem solving, reflecting on problem solving, and then mentoring. Each of these elements was developed in a particular way. In the first part of the mentoring guide, pre-service teachers were asked to solve a PoW math problem. Figure 6.3 is an example of one of the problems that was used in the guide. After pre-service teachers solved this problem, they were asked to reflect on their own thinking and their own work on the problem. They kept a brief log or diary of those reflections. Next the pre-service teachers were then given some instruction on mentoring and asked to mentor three of their colleagues from their university class who had also solved the problem.

The problem in Figure 6.3 is from the PoW library and it is called seven congruent rectangles. In this version of the problem, there is a specific question we are asked to answer. We are given the area and asked to find the perimeter of the larger rectangle. (The Math Forum has

Math Forum – Demo Problem of the Week

Seven Congruent Rectangles

These seven congruent rectangles form a larger rectangle. If the area of the larger rectangle is 756 units^2, what is its perimeter?

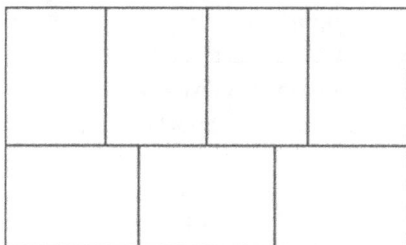

Figure 6.3 Seven Congruent Rectangles Problem in the Online Mentoring Guide

open-ended versions of the same problem that would let the students decide on what was a good question to ask.) Even though the task is specified, there are several ways to solve this problem, including a couple of ways to "guess and check" the problem. By asking the pre-service teachers to not only solve the problem but reflect on their own answers, it set them up to reflect on their colleagues' thinking. They are positioned to engage in reflexive practice, that is, "if this is what I was thinking, what must my colleague be thinking?"

After reflecting on the work of three of their colleagues and posting replies to each of them, the group of pre-service teachers is then asked to engage in a general discussion of what makes good mentoring. This is a guided discussion by a more seasoned mentor, but the pre-service teachers are the ones who develop the discussion and create an outline of the key elements of good mentoring. After they have the discussion on mentoring, they are then shown the Math Forum's guide on good mentoring. They then had another discussion comparing their ideas about what makes good mentoring with the Math Forum's. In the end they attempt to come to a summary consensus of what they think the key elements of good mentoring are for them as a group of pre-service teachers.

The final part of the OMG introduced the pre-service teachers to the back office of the PoW. In the back office, solutions that pupils have submitted are put into their personal cue. The pre-service teachers can then open those solutions from pupils and respond to the pupil with "good mentoring." They would score the students using the PoW scoring rubric we discussed before and then give them a written response to their solution as well. As we said above, there would be an approver, either the student's professor or a staff person, who would make sure the response to the pupil was appropriate before letting it get sent out. There again might be some supportive mentoring here, helping the pre-service teacher write a good response to the student and score them appropriately too.

The OMP was an important step for the Math Forum in terms of developing a service for pre-service teachers. The OMG was thought of as both an interactive textbook and a space where this pre-field experience could take place. As such it was an important step in the transformation of the social space, attempting to make it more conducive for pre-service teachers to meet and think together. The OMG did seem to have a major impact on pre-service teachers' identity and sense of agency. They felt much more confident about math and that they could teach math. Their interest was piqued because they looked forward to working with the pupils in the PoW.

But in a real way, this increased interest and sense of agency was not matched by the pre-service teachers' practices. Of course, their knowledge or ability to do math had not really changed. The Math Forum had provided an important scaffold and it served to have a real impact on pre-service teachers and their interest in math, but the scaffold was not able to go far enough, understandably, and change pre-service teachers' knowledge base around mathematics (Renninger et al., 2006). The question became what to do with this result. In general, we all felt that the pre-service teachers' interest and increased sense that they could teach math was important. But again, we knew that we could not teach these pre-service teachers math and make up for what they missed in their own education. It was decided that the project needed to develop a more robust module on mathematical thinking, a module focusing on looking at student work and really attempting to interpret that work, and then finally practice at mentoring.

This new project was called the Virtual Fieldwork Sequence (VFS). In this project the OMG was expanded into three full modules. The first module focused on what the Math Forum called "mathematical problem solving." The second module focused on diagnosing student thinking, and the third module focused on teaching pre-service teachers to mentor in the PoW environment and then having them do some actual teaching of PoW students.

The mathematical problem-solving module was really a module to practice the Math Forum dialectic. Pre-service teachers engaged in a mathematical practice. They worked on some math problems. They then shared this work with each other and reflected on what they had done. They then attempted to summarize things that they had learned and reflected on in that practice. When piloting this module began, it was hard to get the pre-service teachers, many of them who were not strong in math, to get away from an answer focus and to a procedure-based focus. So the Math Forum began to change the nature of the problems in the module. Instead of having problems with specific answers, they began to move to what they called the "problem scenario." The scenario was an open-ended problem space, where pre-service teachers could develop their own specific problems and then figure out how to answer them. The Math Forum at that time was beginning to explore the use of the problem scenario in a number of arenas to stimulate more discussion and more "process" around doing math, rather than going right to the correct procedures to come up with correct answers.

The problem scenario does not teach math, but encourages an attitude toward math that is more process oriented and requires thinking, discussing, and exploration, rather than just coming up with a formula and an answer. Problem scenarios give students some ideas about a situation, for example, a farmer wants to make fences, has so much fencing, what do you think about this? But there is not a specific area problem or perimeter problem. It lets the student think about what kinds of problems could be created and what they might want to do to solve the problem they created. Something interesting happens when students are given a scenario. They can do a lot more mathematical work on their own and a lot more mathematical thinking. They can invent their own ways of exploring, and develop their own metaphors for the expression of new ideas. Of course, this work needs to be scaffolded too. But in terms of lowering the barrier toward developing their own interest in mathematics, the problem scenario does some important work.

The mathematical problem-solving module had several goals that were in line with reform ideas. First, there was how to help the pre-service teachers develop a sense of agency in the process of mathematical problem solving. The problem scenario was designed to scaffold that agentic process. The second goal was to help the pre-service teachers improve their mathematical communication skills. Problem solving and communication are the basic elements of developing mathematical knowledge. Supporting pre-service teachers to design problems and communicate about those problems helped the pre-service teachers see the connections between rudimentary math concepts and more sophisticated formal methods. It allowed them to see what was going on with formal procedures without being too intimidating and without forcing them to apply those procedures without understanding. All of this could lead to developing productive mathematical habits of mind and problem-solving heuristics, which then could lead to recognizing the pedagogical value of a range in student work, not just what the student got right or wrong.

Because the Math Forum is not rooted in classrooms and does not have a specific curriculum to teach in a timeframe, it's easier for its members to think about exploring mathematics rather than teaching a set of ideas. But at the same time, the insights gained from being able to explore mathematics more freely could be very valuable for classroom teachers. The mathematical thinking module of the VFS project was used in university classes, as a virtual interactive text, which allowed the temporal flexibility of the online to be molded into the time-constrained university class. This is often how Math Forum worked with K-12 classes too. It's not just that the services it provided were an extra-credit activity on the side; it could be central to teaching an attitude or stance toward mathematics while allowing the teacher to remain tied to the temporal yoke of the curriculum.

Following the Math Forum process of engaging in the practice, reflecting on one's own practice, looking at the work of others, then reflecting on that work, this double reflection is reproduced in the modules. In module one, mathematical thinking, doing the work, and reflection on the work is accomplished. In the second module, the second reflection takes place. Here, pre-service teachers were invited to look at pupils' responses in the PoW archive and then to reflect on what the pupils were doing. This module was "diagnosing" the thinking of pupils from the evidence of the work they had produced in the PoW. The module was structured to support pre-service teachers to develop expertise in analyzing student work and to begin to make better interpretations of what the pupils might be thinking. This then would lead to supporting the pre-service teachers' ability to differentiate and adapt instruction based on an understanding of a given student's concept development. Finally, this double reflection process could continue the development of mathematical habits of mind that support the effective use of student thinking as an instructional tool for the pre-service teachers.

The third module was similar to the second half of the original online mentoring guide. In VFS the third module was called Facilitating and Scaffolding Mathematical Understanding. In it the pre-service teachers once again began with problem solving. They solved the problem that they were about to mentor pupils on the PoW on. They then learned some effective mentoring skills based on Math Forum mentoring guidelines, learned how to use the PoW scoring rubric, and give good feedback. They practiced with some canned solutions and then finally began to teach actual pupils in classrooms. They worked with an approver (who was a Math Forum staff person or their own professor) who mentored them as they worked to send out responses to pupils in the PoW.

VFS had a real impact on pre-service teachers, especially pre-service teachers who were not strong and not confident in mathematics. But it also had some problems, too: the tremendous effort to scaffold students and mentor them through the process of working with K-12 pupils ended up alienating some of the pre-service teachers. Some felt it took too long to get to the live pupils with whom they longed to work. And also some of them felt that the pupils did not dialogue enough with them. They wanted more of a back and forth conversation than they were getting in the process.

Mentoring with In-Service Teachers

As well as mentoring pre-service teachers in math, pedagogy and the use of technology, the Math Forum worked on these issues with in-service teachers too. In fact, one could argue that all of their projects with in-service teachers involved a significant amount of mentoring of those teachers in math, pedagogy, and use of technology. One important project, where mentoring was at the core of the project, was a National Science Digital Library project called Leadership Development for Technology Integration (LDTI). The idea of the project was to support a core group of teachers, who would then become leaders in the integration of technology and technology tools to advance teachers' understanding of mathematics. The lead teachers went through online training and face-to-face workshops. They then took this training to their schools and districts where they supported others in developing these competencies. There were also a couple of groups who were working on developing new online resources for teachers to use to enhance their own understanding of math and their work with students.

The core instruction around this project was a six-week module that the Math Forum did simply on their web server and not in a course management system called Algebraic Reasoning. The core of the module was to explore algebraic reasoning through the use of the Math Forum's collection of technology PoWs, or tPoWs. Further, the project took advantage of other tools in the Math Tools library, the new library the Math Forum built as an effort to expand the Math Forum community in new directions

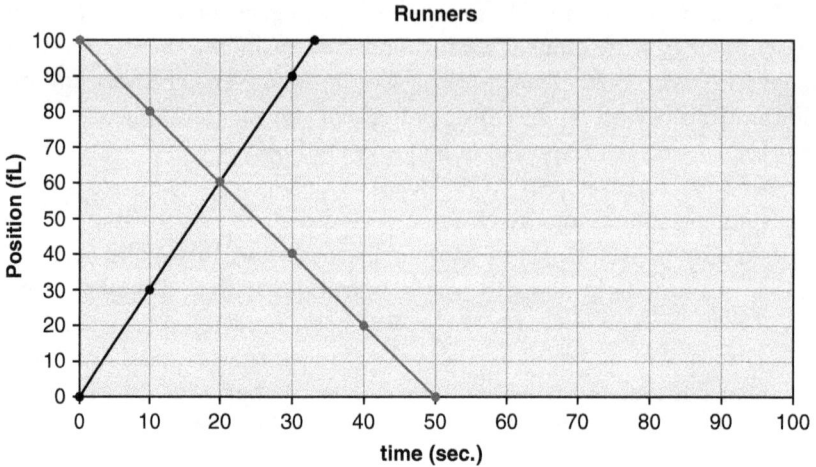

Figure 6.4 Runner's Graph Created Using NCTM 1998 Online Applet

and create a new subcommunity. The algebraic reasoning module begins
with what they called the runner's graph (see Figure 6.4). This is a graphic
from the NCTM. In Chapter 5 we talked about how I compared air travel
speed and head wind and tail wind to running uphill and downhill and how
that made its way into Math Monday. The runner's graph depicts something similar. It shows two lines representing two runners, one at 100 going
to 0 and one at 0 going to 100 and the time it takes them to go from start to
finish or finish to start. Time is the *x*-axis and position is the *y*-axis. The
graph is meant to stimulate a conversation about what is being represented,
what information can be obtained from the graph, and what it can tell us.

It is interesting to note how many people, skilled teachers and novices
alike, first see the graph as representing running uphill and downhill. But,
in a way, this mistake sets up a wonderful opportunity for conversation.
Some individuals catch their mistake right away and are often amused.
Others take time to see that graphs do not need to be indexical signs.
Rather, graphs are symbols where there is a conventional way of producing
meanings that can then be interpreted from the graph (Peirce 1982). This
mistake then scaffolds a close reading of the graph. Teachers are then
naturally predisposed to ask the question, What else can we see if we look
closely at the graph? This then yields a detailed conversation about the
relationship of the graph to specific kinds of information that it does
display. But it also naturally starts the reflection process; if some teachers
misinterpreted the graph, many students are likely to do so as well. And this

then led in many groups to a pedagogic discussion of how one could support students looking carefully at the graph and discussing what might be seen in the graph.

The next phase of the module moves on to looking at an applet that also depicts two runners moving in opposite directions from 0 to 100 and from 100 to 0. The applet is similar to the graph with the exception that because it is an applet you can vary the step size of each runner, where each runner starts, and whether they are going in the same direction or opposite directions. The teachers are then encouraged to play with the app, look at the graphs that the different configurations would generate, and then think about how their students might work with or think about this. From this point the module offers teachers to options of where to go next. First it has the teachers go into the Math Tools library and explore the tPoWs and other applets in the library. The other direction they can go in is that after they have done some exploration, they can solve some problems together, using tPoWs, and discuss that work and what it says to them about algebraic reasoning.

The summary of the module is then to reflect on what it means to engage in algebraic reasoning and to reflect on common misconceptions that student have and how one might address them. In the module itself toward the end of the course, the Math Forum staff state:

> Alan Schoenfeld, a mathematics educator, calls problems "starting points for serious explorations, rather than tasks to be completed" (Schoenfeld, 1992). As you work problems, are there things that you wonder about or want to think more about? Do you see connected problems to explore? That's how mathematicians work (Malkevitch, 2003). When students suggest topics in which they are interested, and explore those, it can lead them to take responsibility for their own learning. How can we encourage students to take such an approach to problem solving?

In general, the module has a wiki-like structure. It is not a wiki, it's a simply web page that staff members can modify on the fly. But the simple wiki-like structure makes the six-week module appear to be more of a workspace and less of an online course the way that CMSs tend to make courses look. And connected to the module are actual wiki workspaces for each of the participants so that they can post their ideas and some work they have done and allow people to comment on what they have done. Some of the teachers who did the online module also attended summer workshops where they continued this work by developing plans to lead their own modules at their institutions and made plans to develop new modules.

LDTI focused on the notion of leadership development. Leadership development was supported in a number of different ways. First, teachers were scaffolded to be leaders in their own classes where they focused on not just teaching algebra but mentoring students into the "habits of mind" that we might call algebraic reasoning. Second, a group of these teachers was mentored by the Math Forum staff through face-to-face workshops and online follow-up activities to take a greater leadership role in their schools and communities. It has always been the Math Forum's belief that teachers can be leaders in supporting others to focus on engaged mathematical practices and the related discussions that surround those explorations. Technology played a major role in scaffolding the teachers' work in this project. As we saw in the runner's graph, it created the opportunity to reflect on misconceptions and how others might think about what they see. In the same way, the tPoWs and the wiki workspaces supported collaborations and interaction around mathematical problem solving and mathematical discussions.

Conclusion

Throughout this chapter we have been using feedback and mentoring as similar terms. For the Math Forum, these concepts are about scaffolding individuals in ways that provide enough support to keep working, but withdraw from the individual when they are ready. The support is appropriate, timely, and focused (Pea 2004). We can think of feedback as the particular forms of discourse that individuals might receive, for example, feedback in the PoW environment or feedback a teacher might receive in one of the LDTI modules from either a colleague or a Math Forum staff person. Mentoring is the larger framework into which individual instances of feedback fit. Mentoring for the Math Forum staff is a model of teaching. Feedback can also come from a peer, such as we saw in the OMP and VFS projects where pre-service teachers gave each other feedback on their problem solving. Mentoring does imply a hierarchy and a directional flow. It implies that one person has greater knowledge in an area and they are helping to scaffold others in that activity or way of thinking. But the hierarchy is relative and not absolute. We could see it as positional rather than a role (Davies and Harré, 1990; Harré and Van Langenhove, 1991). While a teacher strives to be a mentor for his or her students, that teacher may also support the students helping each other. And the teacher might be mentored by other teachers, or even their students, in areas where the students have expertise. Mentoring is a way of carrying out a more processual form of learning and teaching.

Just as Lave and Wenger (1991) stated in *Situated Learning* that apprentice groups were not schools, so in the same way the Math Forum is separate from the pressures of schools and the demands of the curriculum where the key disciplinarian of teachers is time. That luxury of focusing on the learning process and remaining focused on problem solving, discussion, reflection, thinking, and then repeating has allowed the Math Forum to explore some important things about teaching that can be applicable in the classroom context. Most important, we also see in the literature the idea of reflexive practice. The Math Forum staff members live the model of double reflection – constantly reflecting on my own practice, which may give me insights into the practices of another as I reflect on their process. At the very least, it gives me ideas about the kinds of questions I might ask, which may stimulate conversations where we build knowledge together as we engage in intersubjective meaning making (Bruner, 1996).

Technology tools like apps and internet spaces in which to work and carry on discussions create more opportunities to enrich the double reflection process. Wenger (1998) in his discussions of how organization can effectively produce communities of practice suggests there needs to be a balance between reification and participation. (I like to also think about participation as interaction, because it's not just a collaborative process but also one where the actions of one have impact on the actions of others.) Technology can be used very effectively to balance this process of participation/interaction and reification. Of course, technology tools constantly create reifications of the things we do and say online. But they also open up more channels of interaction, which as we said before have greater space/time flexibility. This unique mixture of space/time warping with a constant record, sometimes of multiple types, give the reflection and mentoring process a new twist that can bring into the pressured space of the classroom more opportunities for reflective engagement and collaborative learning.

Technology tools can also help scaffold teachers as they attempt to engage in this double reflection process that is part of mentoring. The "guide on the side, rather than the sage on the stage" is a cute and easy slogan. But in fact it is hard work, requires more confidence and knowledge, and implies a more collaborative and reflective work process than many teachers understand how to engage in. So in other words, it is empty rhetoric. The Math Forum practice of double reflection and using technology to scaffold that process moves that expression from rhetoric to reality. Beginning in about 2008, the Math Forum began to call this scaffolding process "Noticing & Wondering." It's a reification itself that was designed to ease the process of moving into a more reflexive position. It is to that concept and some of the ways that concept developed that we now turn.

7   Noticing & Wondering in a Mediated Environment

Introduction

This chapter looks at how the notion of Noticing & Wondering (N&W) developed out of the practices at the Math Forum. The chapter opens by going back to the Math Forum culture discussed in Chapter 1 and looks at how N&W developed over time. The chapter explores how N&W moved from a teaching strategy to become a much more powerful idea at the Math Forum (Ray-Riek, 2013), and further, how it came to consciousness among Math Forum staff slowly and over time, even after the terms were already being used. I will then look at how N&W links to the literature, specifically the literature on Teacher Professional Noticing (TPN) and Mathematical Content Knowledge for Teaching (MKT), in order to see how N&W fits within the larger contexts of ideas in math education. Finally, I will look at the significance of N&W for teachers and how it has worked as a scaffold for teacher practice in Math Forum activities, especially in one recent project. The chapter concludes by looking at the potential significance of N&W for the fields of math education.

Math Forum Culture

Chapter 1 was about how the Math Forum culture developed from three distinct strands. It grew out of the culture of an elite small liberal arts college, the early utopian culture of the internet, and the personalities of its founding members. I discussed how the founding members' love of mathematics, their resilience as problem solvers, and their excitement for computers and internet technologies shaped the early practices of this group. I also talked about how interaction from the beginning was hybrid, blending online practices with those that are face to face. Here we will discuss in a little more detail how this early culture developed over the years. The Math Forum began as a group committed to the idea of doing math as a part of everyday life. But over time it became even more committed to the everydayness of mathematics. This everydayness was instituted in members' own practice, with activities such as Math Monday, and they worked on making sure that the problems they created for services such as the PoW and other activities they did with teachers and students were based in the real world and encouraged an everyday mathematical focus.

The everydayness of mathematics is linked to their belief that everyone can do mathematics and that problems are best when they are real world problems. These are problems that everyone can relate to and maybe even realize they have tried to solve (even if ineffectively) in their own daily practice. Not only can everyone do math, but everyone needs math and needs to do math as part of their daily lives. It's a necessary form of literacy. Because everyone needs to be able to do math, developing the practice of regularly engaging in problem solving and deep thinking about mathematics is not just useful for engineers and other specialists, but can be a benefit to all. Of course, doing well in math is also an excellent predictor of how well one might do in a number of other areas because of the reasoning skills that get developed while doing mathematics.

So as we have seen several times now, Math Forum culture is rooted in values that are certainly consonant with reform mathematics and social learning theories. I've suggested that the dialectic for the Math Forum is a movement between problem solving, discussions about that work, and the knowledge/learning that is generated out of that practice. Math is a process in which one is always engaged. Sometimes, as a short hand the Math Forum has referred to this as "mathematical thinking." But mathematical thinking is a process that, rather like Sfard's notion, is communication and cognition as a social process together.

In order to encourage people to share in this practice of doing and thinking about math, the Math Forum has found it valuable to engage in several strategies of disruption. Disruptions are both about disrupting the assumed social norms that one might find in the classroom and disrupting the processes by which people go about problem solving, especially when they are following procedural routines. And as we saw in the previous chapters, one of the key elements of that disruption is to encourage a process of reflection among both teachers and students.[1] But the kind of reflective practice that the Math Forum encourages is challenging, especially for teachers. It takes real skill to work with students and to really have a conversation with them engaging their interests and ideas. And many schools, administrators, and even some teachers fear this approach is a waste of time. Encouraging reflexive practice among teachers is complicated because it involves double reflection –a complex form of reflexivity. First, I must think about what I was thinking about while I was doing the problem, and then I can move to how this influences my thinking about what my student is doing and how I should connect with those students. Reflexivity, in the form of this double reflection, became a key element of Math Forum culture that was looked at from many different angles. Further, how technology can facilitate the process of double reflection became a key concern of the culture as well.

Noticing & Wondering Comes to Consciousness

In several of the Math Forum projects working with teachers, and pre-service teachers, it was found that there was a tendency to talk across each other. Even though the Math Forum staff and the teachers involved in the project would use the same words, they did not tend to mean the same thing. For the classroom teachers, as well as university faculty teaching pre-service teachers, and the pre-service teachers themselves, the focus was always on the content of the lesson and adhering to curricular agendas. The questions for these teachers were ones such as: What math have they learned? How well have they learned it? What evidence is there that they learned the math? How can we scaffold them better to learn the math? and What are the best practices or best processes for moving this learning forward? While these teachers were often idealistic people, the discourse they engaged in had several ideological constraints. First, as Judith Butler (1997) would suggest, it is a discourse that hails individuals as agentic subjects. And in that it hails individuals as free agents, it hails them as either good at math or bad at math. So it reduces the interaction to a measurable item in the heads of individuals and then potentially produces an alienation or resistance to finding out if they successfully met the norms or if they failed (Cobb et al., 2009). Second, the discourse focuses on successful mastery of some reified assumed preordained curriculum or skill. So the ideology has a reduced vision of the individuals involved and the activity in which they were engaged. And it should be pointed out, since this is an element of the normative culture, the well-intentioned individuals involved have little idea of the limitations of their discourse or that they were working on something a bit different from what the Math Forum wanted to work on.

The Math Forum, on the other hand, as we have been saying in the last couple of chapters, has been most interested in the social process. It is interested in the process of doing math together and related activities, such as talking and thinking about what was going on, in the effort to make sense out of the problem, the paths that individuals take to problem solve, and, importantly, what emerges from the groups' interactions. The Math Forum staff was engaging in what we began to call "taking students' ideas seriously." And, of course, what is important about taking students' ideas seriously is that one must enter into a genuine dialogue with them. So there was a kind of double difficulty here. First, the reflexive position is difficult because it requires thinking. It requires that one become comfortable with one's limitations (mathematical or otherwise) and open oneself to a genuine dialogue. But further, because we are ideologically trapped in the belief that math learning is going on primarily in the heads of the individuals and that what is going on is that some reified content is getting lodged in neurons, then it's harder to even recognize learning by doing and the intersubjective meaning-making social nature of learning. Over time, the Math Forum, attempting to work across this cognitive gulf between its own worldview and the worldview of teachers, began to hit on a scaffold for managing this difficulty that came to be known as Noticing & Wondering (N&W). And I should emphasize that it was a slow process to realize that the ways that teachers thought about math education, tied up in their practice as it was, and the way the Math Forum thought about doing math were different.

N&W literally began as a heuristic device. In workshops Math Forum staff might ask the students with whom they were working, or the teachers with whom they were working, what do you notice, and what do you wonder? In student work, it begins the process of talking about what the student sees in the problem. A student can notice things about the information provided, the kinds of units or measures, the questions they are being asked, and so on. And if students have trouble noticing things, they can be prompted. In the same way, they can be asked, What questions do you have? in the form of: What do you wonder about? A student could wonder about the appropriate solution path, if some way of working will work, if there might be an order to responding to the problem, and so on. The focus is on what the student is thinking and doing. In the same way when working with teachers, the Math Forum staff can ask them to look at a problem and notice and wonder. They can ask the teacher to compare what the teacher notices and wonders about a problem with what their student notices and wonders. They can also ask the teacher to notice and wonder things about the student's solution. In this way, noticing and wondering supports the double reflection process for teachers. Further, it keeps them rooted in the evidence – what their students are doing. And it keeps people rooted in the mathematical activity that is going on and leads to taking student's ideas seriously.

While N&W goes back to the beginning of the Math Forum's work as an online math education community, and while the concept developed over time, the specific terms developed sometime between 2005 and 2008. Because N&W developed organically and over time, we don't have a precise date for its beginning, but by 2008 staff members were posting short articles on the site about noticing and wondering (Fetter, 2008; Hogan & Alejandre, 2010). Before they actually coined the terms, they began to use the words to notice and to wonder in their conversation with teachers. Over time they began to see that N&W not only could be words to focus teachers on what to pay attention to, or what to get students to pay attention to, but could function as a scaffold, an entry point into learning. And while the Math Forum uses the N&W model in face-to-face workshops, N&W has always been technologically mediated, because working with communication technologies is always part of what the Math Forum does as a math education center on the internet. N&W is not only about what students notice in a problem or what teachers notice in student work, but it is a framework for anyone thinking about math, either their own work or the work of someone else. One teacher could notice and wonder about another teacher's work as well as she could about the work of her own students. Importantly, students are encouraged to notice and wonder about the problem context, their own work, other people's work, and more.

The core of N&W is a way to talk about mathematics and a way to encourage taking mathematical ideas seriously. It is not primarily a rubric of evaluation, but is also a scaffold – a reification – designed to help one stay focused on the moment of the doing of math. The Math Forum staff would say that it's a natural way of learning and that in all arenas of life, this is how we learn. We notice things about the problem or the situation of which we are a part, and then we wonder about how to address that situation or what we can do to alleviate the problem or problems we have here. As we have seen in the last several chapters, for the Math Forum, doing mathematics is at the center of learning to think mathematically. And so one wants to stay genuinely engaged in the work of the doing and not leave that space too quickly. Noticing was the way that the Math Forum scaffolded this first step of reflection. It allows the individual to think about what the problem is, what questions it raises, what possibilities it opens, and maybe what it forecloses. Wondering slows the problem-solving process down. It keeps the individual from doing things too automatically. It disrupts the normative assumptions about how math should be done, what procedures should be used, what a question is actually asking. It forces us to link the moment of the problem to the larger universe of our mathematical understanding. This is mathematical thinking.

N&W is an activity to engage in with students, with teachers, and with one's own mathematical work. Sometimes the staff would do workshops where both students and teachers were involved, and teachers could see how the Math Forum used N&W with their students. In their workshops with teachers, the staff was aware that often we all miss good teaching moments because we are too focused on what we expect to see when students solve problems. There are other things that distract us, too, such as our focus on our own way of understanding a problem, or our failure to see the genuine mathematical thinking that is going on when someone makes a mistake and goes off course with a problem. And historically when the Math Forum encouraged teachers to focus on these moments, it was hard for teachers to understand what they were doing and why they were engaging in this seeming waste of time.[2] N&W was a kind of formula for suggesting that we should try to disrupt these "habits of thought" and try to really focus on what we are seeing in the work that others are doing and what kinds of questions that seeing might raise. As we link the specifics of the problem to our larger knowledge, we increase the likelihood that we might make an important connection, even when the problem solver goes off track.

In the same way they worked with teachers, the Math Forum staff also worked with students and encouraging them to notice and wonder as part of a problem-solving strategy, in particular as a starting point to understand the problem. Again this was a conscious strategy for disrupting students' habits to immediately solve a problem or do mathematics without necessarily making sense of it first. When one is genuinely engaged in the interactive process of problem solving, a strategy to help think about doing the problem is the same strategy that a mentor can employ for thinking about what the problem solver is doing. But if we want to disrupt the ways that teachers think about student work, for students interestingly this can be the way to form good "habits of mind." N&W is a mathematical practice and a natural way of thinking. It encourages the student to not reify math and see it as something different from other kinds of thinking. Rather, one should think about math with the same tools we might use to think about how to do other things. This "good habit" disrupts some of the stereotypes students have about doing math, for example, that math is just about plugging in the right formula. Ironically, N&W is a reification, a formula, that is designed to disrupt formulaic thinking. Of course, it does not always work; sometimes it is simply done as a formula or people notice things that are off topic. But the Math Forum would suggest that is better to try and engage that thinking and move it back on topic than repress it.

Technology tools have been helpful to the Math Forum in this process. A key thing that communication technology can do to slow down the problem-solving and thinking process. For example, students can write in solutions to a problem and submit them through the Problem of the Week (PoW) platform. The PoW platform can encourage them to slow down, explain their work in detail, and think about what they have done before submitting. But the same platform allows the teacher or mentor to slow down their own response to the student. They can think about what they are seeing in the work the student has provided. They can think about what might be a useful intervention. The platform provides a wonderful space for a more robust N&W. Slowing down is one of the important affordances of a technological environment. It's important to note that the slowing down is not an exercise. A teacher could encourage slowing the thinking process in a face-to-face class, and that could be a useful exercise. In the online PoW platform, slowing the thinking and problem-solving process is part of the nature of the environment. No one has to institute it as an activity. The online environment affords other important structures that facilitate N&W and reflection on the mathematics one is doing. It flattens the social hierarchy in ways that again are harder to do face to face. And those flatter social interactions allow students, teachers, and others to talk as social equals, to some extent, in their efforts to understand mathematics. In this way technologically mediated noticing and wondering is a unique variation of the N&W process.

Another key part of N&W was moving from traditional closed-ended math problems, where there would be specific answers that were sought, to open-ended "scenarios" where individuals could explore the problem space and maybe come up with different problems to solve. The Math Forum staff would come to argue that this shift in presenting problems really opened up the power of N&W. As discussed in Chapter 6, the "scenario" approach gives students a situation with a set of constraints and some information. The student then has to explore that situation and develop problems that can then be solved in the situation. Here it becomes important to notice things, then wonder about things and slow down the whole cognitive process, rather than quickly moving toward an answer and potentially missing important insights. Although students can work on scenarios individually, the scenarios naturally lend themselves to collaboration and discussion. Students can share with each other what they noticed in the scenario, what they wondered about, and how that led them to problems they wanted to try and solve. It is a kind of self-scaffold for developing some interest and moving toward a way to solve a problem and learn something new that you did not understand ahead of time.

Further, problem scenarios have the same effect on teachers. Teachers must sit with the scenario themselves and notice and wonder. They must think about the problems they would create out of the scenario and how they might solve those problems. When teachers look at the work of students, they cannot move quickly toward assessing whether the student got the problem right or wrong. They must attend to what the student was thinking, which is facilitated by noticing what the students did with the scenario, the problems they came up with, and where they went with that. Teachers need to wonder about the thought process and what kinds of mathematical ideas the students have (wrong or right) that went into the ways they constructed problems and how they solved those problems.

Theorizing Noticing & Wondering

An important and related discourse to that of N&W is the literature that has developed around Teacher Professional Noticing (TPN). To encourage classroom teachers to stay focused on the evidence of what students are doing, TPN is used as a model for thinking about how to move forward in a classroom where the teacher is watching how students take up ideas, solve problems, and talk about their math work. Jacobs et al. (2010) see this as a three-stage process where teachers attend to what students are doing, next they interpret what is going on with the student, and finally they make decisions about how to intervene or how to move the instruction forward.

Noticing is at the core of the practice of TPN. Noticing is a reflexive process of attending to the behavior of the students in the classroom and then figuring out what are the most important features of that behavior (including discourse) that need to be attended to. It is not just a matter of remaining evidence based, it's a matter of having or developing the professional knowledge to understand what are the important pieces of evidence to attend to. Drawing on the notion of "professional vision," researchers suggest that what develops among mathematics teachers is a practice of how to know what to focus on when noticing the work of students (Goodwin, 1994; Jacobs et al., 2010; Sherin et al., 2011). These professional practices come from experience working in the classroom with students and having a sense of what are the important comments and behaviors, and what are the less important ones. With practice, teachers learn what to attend to.

The interpretation phase of this process is just as important as the professional noticing (Jacobs et al. 2010; Thomas et al. 2015). This is because this is where the teachers theorize about what it is that the students are doing. Noticing and interpreting cannot be completely separated. The teacher knows what to notice, because they already have a theory, and they can further interpret what is going on with the student, once they have attended to a critical moment. Theorizing about what it is that students are doing means focusing on what it is that students are thinking. Drawing on the developmental literature, the researchers in the TPN arena point out that children think about math differently from adults and so have different ideas, but also might make different mistakes due to developmental limitations. Interpretation involves having a good sense of these theories and some experience seeing them in practice in order to interpret what the students are doing.

Finally, deciding how to act is the last stage of this process of reflexive practice. The goal of TPN is to support classroom processes that involve a "theoretical practice." This involves the self-conscious focus on what it is students are doing by attending to the appropriate words and activities, then interpreting that behavior through theory, and finally deciding how to act. Acting is informed by understanding, but it does not imply there is only one way to act (Jacobs et al., 2010). Teachers might decide to respond to a situation in a number of different ways. A key difference between TPN and N&W is that they have different goals. TPN became a way to think about how effective teachers decide to respond to students in the classroom context. Because of that, it has a means/ends focus. The goal is more effective teaching to move students along in a particular curricular path. Like N&W it's rooted in focusing on what students are doing. But TPN moves more quickly toward having the teacher make a decision about how to act in relationship to the work that students have produced. N&W is focused on developing a relationship with the student and on taking that student's ideas seriously. Math Forum staff also encourage an awareness of the developmental level of the students and what problems and words are appropriate for them, but this is in order to foster a better conversation with the students. That means a deeper sense of what the student was thinking about and maybe, for the sake of the conversation with the student, going further with the student's idea to foster a mathematical conversation and a discursive relationship with that student. N&W seeks to escape the ideological trap of the free individual who must perform for a more social notion of engaging in mathematical discourse and producing mathematical meaning together, mentor/teacher and students together.

Finally, while TPN has become somewhat proscriptive, it began in observational research. So researchers in this area first began to look at what teachers do, and to see how those who really had a "professional vision" worked. Then later researchers began to talk about how TPN is an important skill that teachers could learn and use to improve their classroom instruction. There are several things that go into the math teacher's "professional vision" that would make it possible for them to engage in TPN. First, those teachers need a basic understanding of developmental psychology and an ability to know what it is that the pupils they are teaching are capable of understanding. Further, they need to have a sense of the way the age group that they are teaching would approach a problem. Finally, they need math content, but content that is informed by the above situational conditions, and an overview understanding of the math in the area they are teaching in order to fit it into the way their pupils might approach this work.

A real breakthrough in thinking about these issues came from researchers who were working on what they began to call pedagogic content knowledge (PCK). PCK, as several authors have pointed out, is a term that Shulman developed from the realization that content knowledge of a field alone was not enough and that teachers also needed to have pedagogic knowledge in order to make that content knowledge usable and appropriate for students (Ball & Bass, 2000 Shulman, 1986, 1987; Silverman & Thompson, 2008). Later, as researchers began to focus on what it is that teachers do in practice, and they started to look at real classroom needs, they began to develop a related concept of mathematics content for teaching (MKT). Several researchers discuss how MKT grew out of research into PCK (Ball & Bass, 2000; Silverman & Thompson, 2008; Hill, 2010; Hill et al., 2005).

Ball and Bass (2002: 11–12) reported that when they looked at the practices of math teachers in real live contexts, they saw that math content had to be thought of in three particular kinds of ways. First, teachers needed to do a significant amount of mathematical work in teaching and that included designing appropriate assignments and thinking through how to present that work. Second, teachers needed to explain math work in more detail or, as they say, "unpack" mathematics. And this is different from the way mathematicians or scientists work with math, as for them math can be a shorthand or a more concise form of expression. Finally, they suggest that math is connected across a number of different "domains" and that students must work with more than one mathematical concept at a time. The importance of this work is that teachers need specific kinds of mathematical knowledge, and knowledge that maybe is different from what a mathematician or an engineer might need. The MKT, at any level, involves knowledge that will allow teachers to engage in the reflection process that we have been discussing. It allows them to think about what children at this age level can think about, and what kinds of connections between concepts and processes they need to make in order to help kids move forward. And while not explicitly part of the model, MKT allows teachers to make productive use of the mistakes that children make and the limitations of the child's perspective on any particular problem.

Of course, one of the big problems is that many teachers do not have the MKT that might be needed for them to teach. There has been significant effort to measure the MKT that teachers might have, in order to assess the impact of MKT on student learning (Hill, 2010; Hill, Rowan, & Ball, 2005). We can think of the N&W paradigm at the Math Forum as, in part, one way to address the issue of MKT and perhaps the limitation of teachers' MKT. As discussed at several junctures, the Math Forum is an informal community and therefore does not have some of the same pressures of districts and schools that are on teachers and students in those contexts. This allows them to focus more directly on the process of doing math and learning math, without having to worry about other social pressures, such as how much curriculum needs to be covered in a particular period of time. Importantly for teachers, it is also a relatively safe space to share one's weaknesses and strengths in both math and pedagogy. N&W is an important opportunity for teachers to explore their own mathematics and the mathematics of their students without having to worry too much about how good they are at the particular mathematical practice they are engaging in.

EnCoMPASS Project

N&W can be thought of in several ways. It can be thought of as a heuristic for the more complex process of attending to people's mathematical practice and trying to make sense of that practice in order to move people's mathematical understanding forward. It can be thought of as a scaffold for the above process as well. Finally, at a meta-level, N&W is an evolving form of consciousness that developed at the Math Forum. While the Math Forum began using the term N&W somewhere around 2008, it began simply as part of their own teaching. The staff then discovered the power of this scaffold over time, and its potential to lead individuals to a deeper conversation about math, letting the scaffold retreat into the background. One current project at the Math Forum has been a watershed in terms of the raising of N&W to a new level of consciousness among the staff and the researchers at the Math Forum.

Emerging Communities for Mathematical Practices and Assessment (EnCoMPASS; http://mathforum.org/encompass) is a project that has been developing a software tool to create more opportunities for teachers to notice and wonder with their student work. The core of the EnCoM-PASS tool helps teachers to focus on the evidence of their students' work by highlighting things they notice in student work and commenting on the things about which they notice and wonder. They can think about what is going on in the student's mind or how to push their thinking forward. The tool makes it easy for teachers to compile this kind of work and even share it with other teachers if they like. It also helps scaffold the process of mathematical thinking for teachers, as it encourages them to notice and wonder in every problem-solving moment.

But the EnCoMPASS project did not begin with a full-blown understanding of the scaffolding value of N&W. When the project began, it was focused more on the concept of rubrics and helping teachers to develop useful rubrics for the evaluation and assessment of student work. The idea originally was that the software tool would allow teachers to focus on the work their students did and then to categorize that work in terms of the kinds of mistakes pupils made, the ways that pupils tended to think about the math involved in problems, the kinds of things the pupils did well. From this focus on the evidence, the rubrics would emerge and then teachers could talk about what they saw in the data. The software is just part of the project. The work with the software is then brought into dialogue with other teachers in face-to-face workshops and online Math Forum discussions as well as through the Mathtwitterblogosphere (http://mathtwitterblogosphere.weebly.com). This online community of teachers would then develop a really robust discursive community around mathematical thinking and math pedagogy.

Originally, the developers of the project (myself included) thought of the EnCoMPASS software as a kind of Trojan horse, which encouraged teachers to be evidence focused, to talk about the math work their students were doing, to share with each other their thoughts about how to move students forward, and essentially to flex their MKT muscle without them being too focused on their own anxiety about their ability to do mathematical thinking. Two discursive refrains we have heard at the

Math Forum over the years is that teachers are uncomfortable with the knowledge of mathematics at the same time that they don't have time to talk about math with their students. These discursive refrains may be coming from different teachers; occasionally, the same teacher might utter both. But the effect of this discourse is to limit the power of taking students' ideas seriously and limit the kinds of mathematical conversations teachers have with students. The EnCoMPASS tool was designed to help teachers into that conversation, while not really realizing that is where they are going.

EnCoMPASS, like so many Math Forum projects, was also designed to build a community of teachers through a combination of face-to-face and online activities. There have been two summer face-to-face workshops with EnCoMPASS, some hybrid "usability" workshops where some teachers were working in situ with other teachers who were working online, and finally some online courses designed to integrate the ideas of the EnCoMPASS software into professional development activities. During the first summer workshop, the focus was still on the development of rubrics and how the software could help with rubrics and support teachers thinking about their student work. During this first workshop, it was clear that the teachers were having trouble seeing how this software could help them with their work with students, and how it would not slow down their work and hence be a deterrent to what they needed to do. The Math Forum staff saw that the teachers were struggling, and they and the research team for the project, after the workshop was over, began to think about what they could do about this situation.

One of the reasons the workshop had problems was that the software was at an earlier stage of development than we had hoped and teachers were having trouble imagining it in more robust form. But that was not the only problem; even if the software was further developed, we could see teachers were not quite fully valuing what we were trying to do. It was at that point that we all realized that we had fetishized the software itself, and since we were so focused on getting the software to where it needed to be, it had become kind of a surrogate for what the project was about. At that juncture, the Math Forum director refocused the project and suggested that the core of the project was the idea of "taking students' ideas seriously." This was an important shift for us as a group. It moved us back squarely into human concerns and not concerns with things; as Wenger (1998) would say, we reestablished the balance between participation and reification. But further, for the Math Forum, taking students' ideas seriously is much more than just focusing on their ideas about math. It meant really trying to establish a dialogue with students, understanding what they were trying to say, and not correcting that, but engaging in a conversation with them about what they were thinking.

It was at this point we realized that we needed to make the rubric of N&W a core piece of the software itself. We began to build into the categorization system itself, the ability of teachers to highlight a piece of a student's answer to a problem and then to attach a noticing or a wondering to it as a way to annotate this piece of discourse. At the same time that the work of building N&W into the software was going on, the Math Forum held some one- and two-day workshops that were sort of usability tests for the software itself but were also about working with the teacher community to make sure the software and the project met their needs. The Math Forum took seriously the teachers' concerns with wanting the tool to speed up the evaluation process, not slow it down. And while they tried to address teachers' needs around efficiency, they were aware that once people got invested in understanding their students' thinking, they stopped worrying so much about how long it was taking.

By the second summer workshop the teachers were on the same page as the Math Forum staff and the EnCoMPASS researchers. The workshop had a different feel as each group was moving toward how to make the software tool better and how to build the online community so that new teachers could get involved. By this point everyone was talking in great detail about N&W, and it was clear that the N&W paradigm had become a central part of this project's effort to take students' thinking seriously. A number of teachers who have worked with the software more recently feel that it could effectively speed up their analysis of student work. While speeding things up was not a goal, the outcome was positive: teachers felt they could attend to student work carefully and not have that interfere with the pace they needed to keep up with their workload.

While it was clear that the concepts of N&W helped teachers in the EnCoMPASS project see the notion of a rubric differently, it also allowed them to think about taking students' ideas seriously without worrying too much about how much time that would take or how it was not an efficient strategy for teaching math. Of course, they still worried about those issues, but the core group of EnCoMPASS teachers are thinking pretty deeply about teaching mathematics. They have their own blogs on the subject, and the Math Forum community that EnCoMPASS has built has been grafted on to their own communities where they think a lot about these issues.

Significance of Technologically Mediated Noticing & Wondering

N&W then has become a key model for framing and scaffolding the effort to "take students' ideas seriously." As we said above, it is related to notions of TPN and MKT that are found in the math education literature. N&W encodes the Math Forum principles of engaging in a process of double reflection when focusing on students' ideas. One has to always reflect about one's own problem-solving process, for example, What was I thinking? Why did I choose the strategies I chose? Where might I have made mistakes? Then one can go on to reflecting on student work and how it is a "trace" of the thought process of the student. We engage then in an informed process of interpreting what the student was doing and attempting to connect with the student through a reflexive inquiry.

As the Math Forum staff would say, N&W is a natural way to approach a problem. It is the way that humans talk about things and attempt to make sense of things. In a real way it is central to the process of learning or knowledge production as we have come to define it; it is a process of intersubjective meaning making. Because the Math Forum has always been about organic conversations, and how conversation is a central part of learning and knowledge building, the staff naturally began to talk with teachers and students about noticing and wondering. It was from that natural informal process that N&W began to become a model and then a scaffold for giving people a start into a reflexive conversational process.

Importantly, N&W is almost always technologically mediated at the Math Forum (TMN&W). N&W is a powerful heuristic, but when it becomes TMN&W it is even more powerful. And this is one of the important things about internet technology for learning in general. The transformation to TMN&W is a quantum leap, not just additive. As we have seen in other places in this book, because learning and thinking are social ventures that are at their core very much about communication and meaning making, different ways to communicate ideas help to create different contexts for interpretation, and different meanings flow from those different contexts. But it is not just that there are different ways to communicate. The internet literally transforms the social space within which actors do the work that they intend to do. And because all work takes place within spaces of production, the internet makes completely new ways of working and learning possible. In the next two chapters we will talk about some of the ways the internet transforms spaces and identities in the process of making new opportunities for communication. For example, it's not just that Twitter allows people to share 140-character messages with each other. It is that Twitter becomes a specific kind of broadcast medium that allows social groups to organize on the fly through hashtags and can be, as we have seen, the organizing communication system for revolutions that people around the world can participate in.

In the same way, TMN&W opens up new ways to notice and wonder. It allows the slowing down of the interaction process that we have discussed. It allows a teacher or a mentor to express in more detail their own thinking about a problem as they ask students questions about what they were doing. And this process not only inquires into the students' thinking, but it demonstrates, without having to tell them, why it's good to show your work and talk about what you were doing when you are solving a problem. TMN&W opens up new ways that mathematical work can be social. It not only scaffolds students in ways to begin to think about problems, it scaffolds teachers in ways to talk to students about their work and supports a mathematical conversation. Technology allows conversations to proceed slowly and clearly enough that student and teacher can both get a good bit from the interaction. And it helps in the disruption process of normative, procedural practice. The EnCoMPASS software tool specifically evolved to take advantage of TMN&W. It also has the potential to evolve further and be a social platform for N&W and sharing that N&W with others to deepen the conversation that teachers can have about their mathematical process.

Conclusion

As I said at the beginning of this chapter, the Math Forum has, from the beginning, engaged in reflective practice. And it has, from the beginning, been aware that reflection requires that one think about one's own practices in order to make sense of the practices of others. N&W is the mature phase of this process of double reflection. It is a scaffold for teachers and students. But it is a scaffold that is designed to erase itself and lead the interlocutors to a deeper conversation with each other.

For the Math Forum, fostering online communities is about creating a space where like-minded people can talk about math and share resources with each other. It is a place where students can gain new insights into mathematics and where teachers can think about their own math as well as their pedagogy. TMN&W evolved as a wonderful strategy to advance the goals of teachers and students talking and thinking about mathematics together. In the early days, the Math Forum was the locus of these online communities, where talking and thinking about mathematics could go on.

But with more recent projects like the EnCoMPASS project, the Math Forum becomes one node, perhaps a very large node, in the social lives of teachers and students. Many teachers have their own blogs and links to other blogs, participate in the Mathtwitterblogosphere, and find themselves part of online communities like the Math Forum. In this world, N&W finds itself caught up in a complex and different social space where there are many possibilities for producing new knowledge. In the next chapter, we discuss the spatial transformations that the internet makes possible and the evolution that changed social space from the early days of the Math Forum to the current internet world in which the Math Forum finds itself.

Notes

[1] I want to thank my colleague Valerie Klein for helping me focus on the concept of disruption. It's her work with the Math Forum that really brought this concept to the fore.

[2] Ilana Horn in her blog has a wonderful post about wondering. In her example, a teacher was doing N&W with his students, and his administrator felt it was a waste of time. It's a good example of the kind of pressures that are on teachers. It also gives Horn an opportunity to reflect on why it's so important to take students' ideas seriously. https://teachingmathculture.wordpress.com/2015/04/02/what-i-notice-and-wonder-about-teaching-like-a-champion/.

8 Space, Affinity, and Consciousness

Persistence and the Metaphor of Space

Chapter 2 began a discussion about space in ethnography, especially in relation to the digital and online experience. Because ethnography is about understanding people in a particular location, it was necessary to talk about space and online spaces in that chapter. There the main concern was with how to think about the ethnographic enterprise and the ways in which ideas about space and spatial arrangements impacted the process of doing research. Space has continued to be one of the main themes throughout the book. Here I will continue that discussion but now more as an analytic discussion about the things I have seen the Math Forum community do, and the spatial context within which they carried out their work. As such, we can think more deeply about the different ways in which the concept of space has been used by social scientists and how it helps us understand the impact of the Math Forum, as a premier Internet community, on mathematics education and education in general.

It is not surprising that internet researchers picked up on William Gibson's literary metaphor of "cyberspace" to define the new social world that the internet was bringing into being. We had had electronic communication for a long time before the 1980s, but there was relatively little persistence in those electronic communications. One might argue that the phone answering machine was the first mass-market social cyberspace in that it introduced persistence in communication for many people and people could organize part of their words around those messages that persisted. Persistence in electronic communication not only alters the way we move through space; it alters our temporalities. In the age of the telephone answering machine, it was possible to leave a friend a message suggesting that they meet you in a particular location at a particular time.

The friend could remotely log into their machine and pick up this message and make the rendezvous. While this technology altered our practices, our social spaces, and our temporalities, we did not yet think about it as a place.

The development of personal computing devices, TCP/IP, servers, and so on. added to the above story. It's a much more complex answering machine. Now we can have synchronous interactions, asynchronous interactions, the transmission of all sorts of written texts, audio files, video files – the list goes on. And everything persists: materials can be copied, reused, transformed into something else, and so on. A lot more practices, of a complex sort, become possible across very large distances with all sorts of temporalities.

A worldwide web of networked servers that are accessed by computers and other mobile devices has not only made the world small by compressing space and time, but it has created new spaces for human social interaction. Of course, these new spaces are metaphorical. They are indeed spatial, in that they take up some kind of hard drive space and exist on computers and servers that are locked away in some building somewhere in temperature-controlled rooms sucking up vast amounts of electricity. It is interesting that in other developments in electronics, telephony, etc., the image of these tools and the vast literal space they take up have always been images in popular culture. But server farms are rarely seen. If you Google the term, you can bring up nice images of larger server farms, but they do not tend to be images that we see in popular culture. Perhaps this is because they are the "machines behind the screen," so to speak. The paucity of server farm images facilitates the imagination of the vast virtual spaces they open up. The "cloud" or "virtual space" just exists. To actually think of it as a hard drive somewhere reduces our imagination, and the opening of the imagination is part of what is so powerful about cyberspace.

And this is also an issue of reification. If we think about the history of telecommunication from the telegraph to the telephone, we have always needed ways to record the traces of those interactions. Often the trace comes to stand for the interaction. This is certainly the case with cyberspace. No doubt computer-mediated communication makes it possible for much more complex kinds of human interaction, and we can have new social agreements about things like what constitutes a human signature on a piece of paper (itself a trace of an earlier agreement about how to register social interactions). Cyberspace, virtual communities, digital resources are all shorthand for helping us make sense of the new world of possibilities and to engage in efficient interactions in this new world.

In this chapter, we will look at some notions of space that will help us think about some of the things we have seen at the Math Forum. These conceptions of space each draws heavily on a symbolic or imaginative dimension, and certainly the internet has made a tremendous contribution to the symbolic dimensions of space. After looking at the conceptions of space in the next section, we will move on to reflect on how these notions of space help us to understand Math Forum practice and the importance of different conceptualizations of space to the Math Forum. Finally, we will return to some of the work we looked at in Chapter 2 as various researchers have thought about digital media, the internet, and space. This return will allow us to visualize spaces of transformation, and the way that the Math Forum has been a transformative organization for teachers.

Conceptions of Space

In order to think about the "space" that the internet opens up, its useful to think about the different ways social scientists have thought about space. In Chapter 4 we began this discussion in order to think about how the larger society and economy were changing thanks to the new spatial arrangements made possible by the internet. And in Chapter 4, we were focused on the contradictory pressures that both made internet organizations such as the Math Forum possible but at the same time put limiting pressures on them. Here in this chapter we want to focus more on some of the detailed practices that the Math Forum engaged in and the ways that new conceptions of space impacted that work. So we will discuss some of the same theorists, but with a different emphasis here.

One of the most important thinkers about space in the social sciences whom we encountered in Chapters 2 and 4 was Henri Lefebvre. As a Marxist, Lefebvre was aware that under capitalism, those in power organized space in such a way as to facilitate the accumulation of capital. As a philosopher, he was aware that space was not just material but social and symbolic as well (Gottdiener, 1993; Lefebvre, 1991). Capitalism as an economic system goes through constant transformation, as the development of new technologies leads to new products and new processes of production. And therefore the organization of the space of production also goes through constant transformation. Theorists such as David Harvey (2000) and Manuel Castells (1989) have written extensively about the reorganization of cities as new processes of production lead to the reinvention of urban spaces. A number of researchers have written recently about how the old spaces of industrial production have become the new consumer spaces for a postindustrial knowledge economy (Florida, 2005).

Not only does capitalist production get reorganized regularly, but space is also restructured to deal with crises within the economic system. Harvey wrote about flexible accumulation in the 1990s, where he argued that the internet and other communication technologies began to be used to "compress space and time" in order to reorganize companies and use more temporary and part-time workers in order to deal with loss of profit as the postwar economy stagnated in the 1970s and 1980s. This effort to reorganize how corporations did business took off in the 2000s, and much of the productive activity in global cities in the 2000s had to do with the production of services for global corporations by outside suppliers (Sassen, 2001).

If the economic system is constantly reorganizing space to deal with new potential products and means to produce those products, as well as trying to cope with economic crises, Lefebvre was aware that ordinary citizens ended up living in these spaces that the system of production had created. People imagine their lives as they engage in their daily practices within these spaces. People creatively reimagine ways to live in those spaces and use the space for activities other than those of contributing directly to capital accumulation. And so space also has a social dimension as well. The social uses of space have unintended consequences; of course, some of them create pressures for those who are in power.

Harvey (2006) has a related but different three-part division of space. Harvey sees space as absolute, relative, and relational. As a geographer, the absolute is the abstract spaces within which humans inscribe themselves; it's the pure materiality, if you will. Harvey's notion of relative space is that everything within space is relative to other things and relative to the perspective from which all things are being viewed. As Harvey himself suggests, this is sort of an Einsteinian view. Finally, for Harvey, there is relational space. Understanding a point in space or a moment in time is possible only if one understands everything else around it and what is going on all around that moment and place. Harvey suggests two things that come out of this. Our traditional positivistic view of science cannot capture this constantly in motion and relational view, and one cannot think about space without thinking of time too; it is space-time (Harvey, 2006: 121–128).

Harvey then suggests that if we take a dialectical view, each of the perspectives on space-time is in fact in play at all times; there is a dynamic tension among them. He suggests the same kind of process for the Lefebvrian conceptions of space. In order to think about this even further he suggests a model for overlaying his view of space-time with the thought of

Lefebvre. What this does is to give us a different lens on space and space-time to see different kinds of potential in spaces (Harvey, 2006: 135).

An example that might shed some light on this idea that people use the new spaces in surprising ways is Habermas's notion of the rise of the lifeworld in early capitalism. The development of several early consumer industries, particularly newspapers, cafes, and beer halls, created a context where a public could be formed. People could gather to consume these products, but of course they did not just consume, they talked. And the heady combination of information, caffeine, and alcohol created a context for a democratic public debate and spaces for that debate (Habermas, 1989). Habermas calls this the public sphere, and the creation of the public sphere created some dynamic contradictions for both democracy and capitalism for much of the early twentieth century and almost led to the downfall of capitalism.

In some ways the development of mass media resolved the earlier tensions of the democratic sphere in capitalism's favor. But the mass media created new concerns for passivity and an alienated society. By the 1980s many social scientists talked about the death of the public sphere, arguing that mass media and consumer society had created a passive consuming public, no longer concerned with active debate and no longer with the spaces within which to conduct those debates. The ink was barely dry on those social science discussions of the death of the public sphere when the internet revived the discussion of an electronic public sphere. And that is an ongoing discussion because Twitter allows for the organizing of modern revolutions; at the same time, the internet is becoming increasingly used for advertising and consumerism. So is the internet a public sphere for democratic discussion and action, or a consumer space for passive subjects? Right now it is both, and they exist in tension.

Related to Harvey's notion of space-time and his ideas of relative space as well as Lefebvre's notion of lived space and the social and symbolic dimensions of space is the idea of "third space." We also talked about third space briefly in Chapter 4. The concept is used most extensively by Homi Bhabha and Edward Soja. For Bhabha, third space was a way to think about the contradictions of colonialism and how there were spaces for human agency in the cracks, so to speak, of colonial oppression. Bhabha (1994) uses the example of education under colonialism in India. Schools were created to emulate Eton and Harrow because that is what the colonialists wanted. The argument for British colonialism was that the British had a superior culture. But as the South Asians created parallel excellent schools and excellent students and sent many of them to Oxford and Cambridge, this opened a space of anxiety for the empire. The colonies were doing the colonists' culture as well as or better than them, so who was the self and who was the other? What is nice about Bhabha's example is that we see how the physical dimensions of space are overlaid with the lived and the symbolic, the relational and the relative. For Bhabha and Soja third space is this multidimensional space, tied up with the physical geography but also with the social and symbolic geography. It can be a space of transformation and unalienated relationships. It's how "great good" places are possible (Oldenburg, 1989).

For Soja (1996), third space has a past, present, and future as well. His understanding of space is due to his sophisticated understanding of the fact that economic systems in general, and capitalism in particular, are forms of social imagination where those in power needed to lay out a particular imagination in order to exact a moment of wealth accumulation. Further, these systems are always contradictory and so they are initiated, transformed, and then collapse and get rearticulated.[1] The dialectic for Lefebvre revolved around the ways that space was organized to do productive work and then the unintended ways in which people inhabited those spaces and creatively took up new ways of being within those spaces. For Soja, third space is the combination of social imagination and physical location. One might argue that an understanding of the flexibility of social space, and hence social time, identity, and so on, has been enhanced in this contemporary moment where the kinds of space-time compressions that Harvey (1990) talks about have reached ironic levels.

In any event, since our first publication on the Math Forum, Ann Renninger and I (Renninger & Shumar, 2002) were impressed with the ways that teachers leveraged this third space to do things that would have otherwise been hard to do in the physical world. Aware that the Math Forum had helped create a space where social status mattered less, and open conversation mattered more, these teachers felt free to explore new skills and openly ask questions without worrying about appearing ignorant. In this way, before the Math Forum embarked fully on professional development as a service they offered, many teachers leveraged the space to produce informal professional development for themselves, as well as using the resources as a supplement for the students in their classes.

A different way to think about social space is through the work of Pierre Bourdieu. Bourdieu can be thought of as a sort of quantum sociologist who tries to understand the organization of social groups and the ways they are distributed in society. Here space is truly a metaphor and a much more conceptual notion. But at the same time, that conceptual understanding sometimes works its way out in physical spaces too. In the United States, the communities that make up different segments of the middle class often find themselves buying houses in particular neighborhoods, sorted according to zip codes. And then those residents are the parents of the children in the local schools, who inculcate habitus (defined below) into their children. Those children then engage in schooling in particular ways, which may be to their advantage or disadvantage in the educational setting depending on the habitus they have developed.

In general in Bourdieu's model, the social space is defined by different forms of capital. The two dominant forms of capital are culture and economic capital. As a strategy, individuals are focused on accumulating either one or the other form of capital. These strategies are further defined by the social capital, or the connections to other people, one has at one's disposal. And most important in that social capital is the family's position in the class system. In his general spatial model of class relations there are three classes, each defined by levels of social capital: lower, middle, and dominant. Each of those classes is divided into three fragments by the strategies used to accumulate capital: those that are focused on economic capital, those that are focused on cultural capital, and those that are focused on a bit of both.

Each of these class fractions functions like communities of practice. They engage in particular practices of work and social interaction. Those practices are patterned and habitual, and hence become second nature. They are not unconscious, but rather they are below the level of consciousness. These practices are linked to ideas about taste and preferences, from areas as diverse as literature, art, music, food, and so on. These tastes not only shape aesthetic preferences but shape a general approach to the world. If one likes the precision of Bach, then one is likely to favor precision and intellectual complexity in other arenas. Bourdieu calls this habitus. And habitus is inculcated from one generation to the next. It is not just taught, but bodies are shaped by habitus as we are trained in how to eat, listen, sit, and so on. So children from very wealthy parents not only inherit wealth and a social world of connections, but also develop particular ways of being in the world, which are consonant with other members of their class fraction. And they take these attitudes, values, and ways of making choices into the schools where they are educated.

There are many important implications of Bourdieu's model, and the one I would like to talk about here is that while the class system is one of the main social spaces, and perhaps a model for other social spaces, we find different spaces in different contexts, and they will be organized in a parallel but possibly different way. So for instance we can talk about the social space of a university, where faculty are organized along lines of different amounts of economic and cultural capital as well. Further, faculty too have a system of social capital creating something like a class system within the university. Like the general class system, faculty with greater access to economic capital tend to be more politically conservative, and faculty with access to greater amounts of cultural capital tend to be more liberal. But even this is more complicated than that, as biologists tend to be more politically left than engineers, and historians more conservative than sociologists.

Most important to this work, we can think of the world of K-12 teachers as being in multiple and overlapping social spaces. We will talk about this more below, but teachers are part of a middle-class class fraction that would focus on the accumulation of cultural capital, as opposed to say business managers who need cultural capital to get to their position, but they then are more focused on the accumulation of economic capital. As such, teachers find themselves in a set of spaces where some teachers have more cultural capital than others. Higher grade teachers on average have more than lower grade teachers, teachers in affluent districts have more cultural capital than those in poorer districts, and so on. Sometimes for teachers, improving their skills or dealing with pedagogic issues is less than straightforward. In many places, teachers tend to close their doors and keep their heads down. Informal online educational spaces such as the Math Forum can be third spaces in this way too and an important strategy in the effort to increase one's cultural capital and status.

As we have already begun talking about the ways space and time are transformed by digital and internet technology, let's next move to discussing how these theoretical ideas about space presented above can help us understand some of the things we have seen at the Math Forum that we have presented in the preceding chapters.

The Math Forum in Space and Time

The Math Forum has had a sophisticated understanding of social space and the potential of the internet for human emancipation from the beginning. An important aspect to space is that there is some place that persists over time. This notion of persistence is an important early concept for the Math Forum (as well as other research communities and individuals). Space makes it possible not only for things to continue through time, it makes a platform, an environment if you will, within which objects can be developed and changed. An important early object on the internet could be referred to as "text."

Text has received a significant amount of attention in internet studies and research on social media. In the early days of email, there were many stories about people who did not realize the impact of email persistence, and the fact that what one said in email not only could last forever, but could be forwarded to anyone in cyberspace. Before there was the processor speed and upload and download speeds for images, as well as storage space, text was the coin of the realm on the internet. And, of course, text is still central. Text is so central that many researchers have become somewhat confused about how to define research online. Is a discourse analysis of a text-based discussion group ethnography? It's actually a fairly complicated question.

From the beginning, the Math Forum was focused on the persistence of text on the website and the way digital text was a product and a resource for future interactions. These products were originally thought of not as commodities, but rather as community resources. Later, under pressure from their university home to be self-supporting, it turned some of them into digital commodities. But in the early days, these were interactions that were reified by the nature of the digital medium. And as such they could be repurposed and archived for future use. The power of this reuse and repurposing is that the material is still being used.

We have seen in earlier chapters that the core services at the Math Forum, Ask Dr. Math and the Problems of the Week, are very much focused on text, and the repurposing and reuse of that text has been a central activity at the Math Forum. In the early days of the Math Forum, archiving good questions and good answers was a central preoccupation. And making that archive easy to browse or search was also an important concern, which before the days of Google was not a small matter. Ask Dr. Math was a popular early resource. One of the main things that people liked was that there were real, vetted volunteers who were personally answering questions of students, parents, and anyone else who had a math question. Because the volunteer doctors had been trained in the Math Forum way, answers were not alienated bits of math information, but robust conversational answers to people's questions. They were unalienated, even thought they were found in a digital environment.

Even with all this emphasis on live mentors in the early days, what was clear to the staff members was that these interactions could be archived and repurposed as resources for people to benefit from in the future. The Ask Dr. Math staff, even with an army of volunteers, could not keep up with all the questions. So a strong archive was critical. While the Ask Dr. Math tried to answer all questions, it focused on good questions and good answers for the archive. The text was mined, so to speak, for gems. In these early days of reusing text, the Math Forum was aware that this was a community resource and that it could become a valuable asset to the math education community (Shumar, 2009).

The online textual resources of Ask Dr. Math are not just a nice library of answers to common math questions, a math FAQ if you will. They are also discursive snippets that articulate a particular worldview and they echo the culture from which they come. They are in a word, digital traces. If we look at a typical Ask Dr. Math answer to a question, often there will be links to other answers to that same question. The math doctor will begin a conversation around the issue. They will often directly answer the question, but then pick up on a mathematical thread or two that the question and answer implies. In this way a general interest in mathematics is shown, that math is more than just correct answers. Math is a way of thinking and talking about things in our world.

The traces of the PoWs have been used in similar ways. Student answers to problems and the conversations that they have with the mentor who replies to them are also archived. One can look at the archive of the work and see problems in different areas of math (primary, math fundamentals, pre-algebra, algebra, geometry, trigonometry, and calculus), solutions that students provided to those problems, and then the mentoring that we talked about in Chapter 5, including the mentor filling out the scoring rubric and then writing a response to the student. In some cases, we can see a back and forth discussion between the student and the mentor. Over time this has become a massive archive. The Math Forum has built professional development courses and activities out of the PoW archive, including examples of mistakes that students make, how mentors address those mistakes, and so on. These archives continue to be a dynamic living part of the Math Forum culture and community. They are community resources that convey the nature of the Math Forum culture by the way that discourse is structured in the archive.

Interactive Digital Libraries

In 2000 the National Science Foundation established the National Science Digital Library (NSDL). It also had a grant program for NSDL from 2000 until 2011. The digital library movement was an important movement among librarians, information scientists, computer scientists, and others. It was and is a broad movement that began before the web and has continued into the current moment. NSF's NSDL could be viewed as an important part of this general movement in the digital library world. The NSDL portal, which is currently active, provides STEM resources primarily for K-16 education. These digital resources could be books and articles, but also lesson plans, interactive applets, and so on. NSDL also produced a large community of people who developed online resources and communities in the different STEM fields. Math Forum was a central member of that NSDL movement, and the NSDL not only influenced the direction of the Math Forum, but became an important and different community of which the Math Forum was part.

While interactivity was always part of the digital library movement and NSDL's vision of a digital library, understandably there was a lot of emphasis on collections. Here was a virtual space where teachers and school districts, especially those without a lot of resources, could get state-of-the-art materials for their classes, students, and the families of their students. It was seen as having important potential to increase the equity distribution of educational resources. Of course, people were aware that families and schools for poorer areas did not have up-to-date computer equipment or good access to the internet, but those were problems that different groups were working on too. So there was an optimism about how the digital library movement in general and the NSDL in particular could help address resource inequity issues.

As part of that movement the Math Forum has a lot of resources in its collections. And as such it has been an important part of the digital library community for mathematics along with the Mathematical Association of America (MAA), which had the official MathDL in the NSDL. But as the Math Forum was also more of a community or collection of communities, and since the Math Forum spent so much of its time focusing on creating opportunities for mathematical conversations and interactions around mathematics, in the digital library movement the Math Forum discovered themselves as an Interactive Digital Library.

Of course, traditional libraries have been changed by online resources and digital libraries too. In the past, libraries were places of silence where people could engage with scarce resources that could not be found elsewhere. As such they were temples to knowledge. On university campuses they were the center of knowledge and learning, and knowledge production was arranged around them. And as temple spaces, there were strict prohibitions: no talking, no interaction, no food, no drink, and so on. But as the internet has replaced the library as a space for resources, libraries have asked themselves, What are they now? And one common answer, ironically, is that they are spaces to discuss ideas, explore new research, make things with other people. In sum, they are spaces of interactivity.

The Math Forum's Interactive Digital Library similarly is a virtual space of interactivity. Here in a Bakhtian (1986) way, individuals could interact with a range of persistent "already saids" from past conversations and past problem-solving activities, and these persistent traces could be folded into present and future interactions. In the imagination of space, the Math Forum Interactive Digital Library is a social space where people can interact not only with each other but with people who have carried on such interactions before. The historical utterances become part of the current conversation. The Math Forum was aware of this third-space potential and sought to build on it in particular ways.

One way the Math Forum sought to build on the power of the digital library was to create a specific library, Math Tools, within the general Math Forum site. We have talked a little about Math Tools and Technology PoWs in Chapter 6, but here we can talk more about the space of the library. Math Tools is a library of applets and other small pieces of software that are designed to do small mathematical tasks. The software tools can be used independently or built into larger mathematical lessons. The origin of the Math Tools idea grew out of earlier projects where teachers and developers created Technology PoWs, or tPoWs. The tPoWs were problems that had bits of software embedded in them, giving student virtual manipulables to help them think about the math problem they were working on.

The core of Math Tools were the tPoWs, but the Math Forum built relationships with other groups developing small applications as well as other groups who had libraries of applications so that they could all be collected in one space, the Math Tools space. These applets and small applications were made to run on a desktop or laptop computer. This was before smart phones and the mobile revolution. But in a way it was anticipating mobile technology, as these tools were small pieces of code and designed to be mobile. At this point in time mobility meant embedded in other pages and portable so that reuse and reappropriation could be a simple matter. But the front door of the library space was still the web-based home page of Math Tools.

Math Tools was not just a collection of tools; it was also designed to be a community of users, modeled after the Math Forum community in general – a subdivision if you will. There were several pieces to the community interactivity. Community members could review tools, and reviews and tools could be rated by other community members. Lessons could be designed around specific tools and shared on the site. These lessons could be reused but could also be reviewed and rated. Finally, there were areas for discussions around tools and tool use. In typical internet and Math Forum fashion, a lot of the interaction could become resources in the library collection. This was particularly true for lessons designed to go with tools and ideas for how to use tools in classroom situations. Like much of the Math Forum site, many of the participants were teachers, but some were education technology specialists as well. Math Tools has been such a successful digital library space that interaction in the space has continued to this day, even though funding for the project ended in 2005.

Another important digital library project we have talked about in earlier chapters that the Math Forum developed was the Leadership Development for Technology Integration (LDTI) project. An important part of the work the Math Forum does with teachers is formal and informal professional development. Many projects at the Math Forum were not conceptualized as professional development, but became so for the teachers involved. While the first workshops were hybrid professional development opportunities for teachers, over time, the Math Forum became even more sensitive to the potential of a hybrid learning space for professional development. One of the really big problems with professional development for teachers is a space-time problem. Professional development does not allow for the time nor the space to discover what teachers really need, what kind of support would be helpful, or how to provide scaffolds for teachers to advance to the next level of their thinking and practice.

As discussed in Chapter 6, LDTI was a professional development workshop project organized around a core online curriculum in two main areas. The two workshops were titled "Using Technology and Problem Solving to Build Algebraic Reasoning" and "Technology Tools for Thinking and Reasoning about Probability." The workshops were done online, but they were somewhat synchronous in that like an online university course, activities for the first week needed to be done in that week. The Algebraic Reasoning workshop was six weeks long. It was structured much like a Math Forum face-to-face workshop where teachers got a chance to use some tPoWs and think about how those tPoWs could be used to advance their own algebraic reasoning. After reflection on their own mathematical practice, they then got the opportunity to think about using the tPoWs with their students and how to help advance their students' algebraic thinking. There was room for lots of discussion and collaboration. Individuals were encouraged to create their own projects out of the workshop.

The theme of algebraic reasoning was one that a group of Math Forum teachers helped to develop. The teachers and staff were focused on developing a workshop that could be an important and direct benefit to teachers' daily work. There were also several face-to-face workshops for teachers who were interested in the project. The early face-to-face workshops explored the ideas that were put into the online workshops. Later face-to-face workshops were set up to develop new materials for the online workshops. Several teachers involved in the project helped develop materials and also participated in the online workshop multiple times.

Spaces of Transformation

In both of these digital library projects we see that the Math Forum takes advantage of the structure of the internet to transform space and time. That transformation does several things. It opens up the possibility for a third space. This third space is a space of transformation; it not only allows individuals to get at resources they might not have access to otherwise but allows them to think about those resources differently. Math Tools allowed participants to imagine ways to use pieces of software to advance the thinking of their students. Students could engage in the practice of "playing" with virtual buckets of water (for instance) in order to think about issues of volume and measurement and algorithm. This practice could help teachers think differently about lessons and how to structure lessons and stay focused on what students are doing and what students are thinking. Math Tools also indirectly scaffolded this process for teachers. Since it was a community of users who were rating tools, contributing lessons, rating lessons, and engaging in discussions, this was a space where a teacher could contribute to or get ideas from other members of the community.

LDTI was specifically designed to be a space where teacher could engage in professional development. And that professional development had several different paths along which it could unfold. First and most basically, a teacher could participate in the six-week online workshop and get a much more in-depth opportunity to think about their own algebraic or probabilistic reasoning. They then would get the opportunity to use that reflection to help them reflect on work with students and how one might engage students in a mathematical discussion, which would start from where a student was at in their thinking. This kind of in-depth detailed professional development is important for teachers of mathematics. But also, teachers could become part of the community around this work and help to develop new online resources or work to support the community in its activities. This would allow an even deeper connection with the professional development and allow teachers to both produce and consume professional development. Often in this process of producing and consuming with the Math Forum the line becomes blurry, which is a good thing. It helps contribute to creativity and pushing one's thought.

Finally, the internet can be a space where one's status as a teacher and one's strengths and weaknesses can be less important. The Math Forum was particularly aware of taking advantage of this kind of reduced social visibility that the internet made possible. While early research on the internet showed that anonymity increased the possibility for deviant social behavior, a reduced visibility meant the people were still known, but they were known as individuals rather than as statuses. Following Bourdieu's notion of the social space, we can think of every school as a field where people find themselves structured in that space. In that social field, teachers have differing amounts of social power depending on where they stand in the field.

One teacher I talked with long ago had to teach a geometry class in her school. She was certified as a secondary math teacher, but had spent most of her career up to that point teaching algebra. This teacher was not confident about her ability to teach geometry, or at least to do a good job with the class. But she was also not comfortable with sharing her unease with her administration or her colleagues. She felt admitting this limitation would be stigmatizing in the school. Her lack of cultural capital marked her as a less capable teacher. There could very well have been gender dimensions to this tension she felt, but I was at too much of a distance from the school context she was in to get a sense of that level of pressure. This teacher was a part of the Math Forum community, and so in various discussion groups she shared her concern about teaching the geometry class. In sharing this information, she got a lot of support from other teachers online. People pointed her to resources she could use to refresh her own geometry knowledge as well as providing lesson ideas for her students. In a very real way this teacher crafted a personal professional development program that helped her get up to speed to teach this geometry class.

There are a couple of things to say about this story, which perhaps is not an unusual one. But when I talked to this teacher, this was a fairly novel and creative response to a problem. And I talked to a number of teachers who, in different ways, were doing the same thing, crafting their own professional development to meet the needs they had in the school contexts. But here is the most important piece of this story, and why it still has relevance today. The internet did not make this teacher anonymous. But it did reduce the visibility or importance of her status in relationship to the people who helped her. This is what allowed the admission of weakness and the help she got to not be stigmatizing. And this happened because there was a warping of the social field, the social space in which this interaction took place. We can see echoes of Harvey's thinking about the dialectics of space-time here. In her turning to a different community of colleagues, it was possible for them to see this concern about her geometry skills, and the request for help, as reasonable and something that one might imagine oneself needing at some point. But it also possible to imagine that the very teachers who helped online might be more reluctant or unable to do so on their own school contexts, because those social spaces are structured through a complex set of power relations and contradictions. Of course, on one level this is no different from the pre-internet days where one might turn to other friends not in one's school to help out. But the internet allowed for this interactive digital library space of the Math Forum to be a whole, resourced community of support.

There are also limitations to the transformation of space that the internet allows of which the Math Forum creatively took advantage. Another teacher I talked to was able to marshal, through connections she had with the administration and her own creative thinking, the school's computer lab and related resources to do a lot of creative things with computers and the internet in her teaching using Math Forum resources. She had also creatively used the Math Forum space to develop her own teaching resources for her students as well as sharing those resources with other teachers. Fortunately for her, she was one of the only teachers interested in using these resources, and thanks to her connection to the Math Forum and its community, she had a lot of ideas about how to use those resources and improve opportunities for her students to learn math. But interestingly, this was not scalable at her school. If other teachers had learned from her and shown interest in these resources, it would have meant diminished access and opportunity for her. It's hard to say what would have happened if there was a clamoring of teachers asking for more internet resources; maybe it would have transformed the school. But in this situation the physically limited lab space put a limit on the potential of virtual space to be transformative.

Communities, Affinity, and Self-Organizing

In this last section of this chapter, perhaps we should return to the discussion of the anthropologists we encountered in Chapter 2 who talked about the ways that the internet and digital media transformed social space and the site of the "field" for ethnographers. As I pointed out at the beginning of this chapter, the digital's potential for creating persistence in certain media of communication and the use, reuse, and transformation of that media has meant a very real transformation of social space. This is the thing that Miller and Slater (2000), Boellstorff (2008) and others have recognized. Miller and Slater (2000) were aware that social space was constructed out of an interplay of what Anderson (1991) would call the process of social imagination and the technologies that are embedded in the practices that are part of that imagination. As Harvey (2006) would say, this is a dialectical process. Being part of a real everyday Trinidadian culture and community in Miller and Slater's (2000) research has changed, thanks to the potential of the internet. And as Boellstorff (2008) pointed out, it is not just the persistence of digital media that has made this possible, but what he calls the "gap" between what is virtual and what is actual. That gap has always been part of an imagined community, but now as the virtual persists, there are more creative things individuals can do thinking, working, and playing across that gap.

In this same vein, throughout this book I have referred to the Math Forum as a community of practice. I think that is a reasonable model for describing the Math Forum and its work. In the early 2000s after Lave and Wenger (1991) published *Situated Learning* and after Wenger (1998) published *Communities of Practice*, many educators started to use the term in relation to schools and education. There was a lot of concern to create communities of practice or identify communities of practice in schools. Often in my writing I have pointed to Schlager and Fusco's (2004) critique that, in fact, in schools there are really not many effective communities of practice. But the Math Forum took advantage of the "gap" and brought together teachers who were interested in a practice of doing math together, talking about that work, and intersubjectively moving their knowledge about mathematics forward. And this work was not just to advance their own mathematical practice, but to make them better teachers. So the double reflection leads to thinking about how kids think about math and what can be done to mentor them and push their thinking forward. They built an effective and dynamic community of practice in this virtual space, this third space if you will. The space not only made it possible for teachers, across districts, to find each other and share a concern for a deeper understanding of math and learning, but also allowed them a context within which to deepen their relationships with each other. This greater depth was something that was not really possible for most teachers in the pre-internet era.

This virtual third space makes it possible for some teachers to escape the symbolic inequality and potential symbolic violence (Bourdieu & Wacquant, 1992) they might experience in their schools, whether it's gendered inequality or inequality based around class-based differences in cultural capital. But that virtual space is not capable of completely mitigating these inequalities. Many teachers of urban and poorer districts have too many hurdles to overcome to manage the technology needed to be part of this space. Even when the computers and internet are there, there are myriad social forces that might take those technologies apart or make it difficult for a teacher to use them or find the time to use them. And engaging in the more prolonged interaction that leads to the deeper relationships mentioned above is something that is harder when the resources are more scarce.

The Math Forum then is very much a virtual community of practice. But is it one community, or more than one? I think the answer to that question is both. There is an overall culture and general sense of belonging that the Math Forum has been able to sustain over many years. But as I have discussed in this chapter, there are specific communities that the Math Forum developed and that have sprung up on the Math Forum site. Math Tools, for example, is a community that remains active; it is both its own separate group and part of the general community of the Math Forum. Some individuals see themselves as part of only one of these Math Forum subgroups. Other people see themselves as involved with the Math Forum across these different subcommunities and projects. I would argue most teachers see themselves as more involved with the Math Forum in general, and it is the Math Forum culture that they feel very much a part of.

In 2004 James Gee coined the term "affinity space" as perhaps a better term for the kinds of sociality that one sees online, rather than community of practice (Gee, 2004, 2005). For Gee, learning spaces are where individuals share a sense of interest and purpose, and a desire to make meaning together is the important way to think about space when thinking about learning spaces. And communities have many more structural elements, like boundaries, hierarchies, and internal structures. Gee's critique of communities of practice, at least with regard to schools, is similar to that of Schlager and Fusco (2004). I think Gee is quite right when thinking about schools and online game spaces or social media. The online spaces in particular encourage individuals with some sense of "affinity" to congregate in that virtual space and engage in some task with each other (Gee, 2005). While I think that the concept of a community of practice is still very useful for defining the space and the structure of activity at the Math Forum, affinity is also a useful notion as well, especially in the era of web 2.0. In the current era there are more online spaces where there are shared values, and teachers who are concerned with math education can find a shared set of interests; further, there is more potential for self-organizing as communities become more rhizomatic and less centered (Deleuze and Guattari, 1987).

Web 2.0 actually introduces some important new dimensions to thinking about the ways that the Math Forum space works. In the early days and early project of the Math Forum, the virtual space made possible by the internet and digital media was primarily computer and web focused. Websites were the portal into the virtual world, and so the Math Forum was a place people went to. This is still true and the Math Forum's own activity is primarily web based. But today, core members of the Math Forum culture have their own dynamic math blogs; they are active users of Twitter and often members of the Mathtwitterblogosphere (http://mathtwitterblogosp here.weebly.com). The Math Forum now finds itself as a core cell in a rhizomatic community or set of communities that link blogs, tweets, and the online community spaces of the Math Forum. There are now many resources in a number of spaces, and conversations that cross media, between discussions on the Math Forum, Twitter, and in individual blogs. The Math Forum in this context is able to help create focused resources and focused professional development that takes advantage of the third space potential of this hybrid social world. But there are now many more people and groups who share an affinity for being concerned with the everydayness of mathematics, good mathematical interactions, and thinking mathematically.

Note

[1] We could now enter into a complex discussion of how this articulates with Schumpeter's (1961) notion of a "run of capitalism," Harvey's (1990, 2000) powerful important ideas about capitalism and social space, Bhabha's (1994) ideas about third space and hybridity, Hardt and Negri's (2000) ideas about empire, and so on.

9   Identity and Online Interaction

Identity, the Self, and the Internet

If you had asked an anthropologist in the 1990s and early 2000s what they studied, nine out of ten would have told you "identity." Of course, this would be in a mix of other things: postmodernism, globalization, and more recently neoliberalism. But identity was a large part of the field of cultural anthropology. And anthropologists from different schools of thought would have all said they studied identity, but they would have meant different things by that term.

Similarly, as research on the internet began, many internet researchers also focused on identity in their work. From the early work of Turkle in *The Second Self* (1984) and *Life on the Screen* (1995) to more recent work on the self in relation to different forms of social media, identity and the self have been important themes for research too. Ching and Foley (2012) in their edited volume on self in digital worlds adroitly pull together current research in education that focuses on the self and identity in relation to digital media. They also do an excellent job situating the research on identity in digital media within the different frames of research that are salient to researchers in this area, for example, practice theory, situated learning, and activity theory.

What Ching and Foley (2012) open us up to is that scholars in these related areas use a set of terms when thinking about individuals in social groups and the interactions and activities they engage in. Some of these terms are ones we have already discussed: practices, habitus, identity, community, community of practice, and so on. Social scientists are trying to make sense of these terms and articulate a perspective on the actions of individuals in social groups. These terms not only describe the group and the way that group is organized, but also how people see themselves in these collectivities, what they are supported to do, and what they do in spite of pressure not to do those things. So one cannot think about identity without thinking about the related terms of the self and the community or social group – even if that group is just a collection of people who share an affinity for something. And these interrelated concepts necessarily lead to questions of structure and agency, reification and process.

Because there are so many different perspectives on identity and the self in anthropology, psychology, education, and internet research, it would be a large volume to summarize that work. And it is not my intention here to summarize this work. Rather, I will use some of this work in these related fields to help us make sense of the work we have seen at the Math Forum. Much of the work in education that comes from a social constructionist perspective is valuable for thinking about what we have seen in the work at the Math Forum (Boaler & Greeno, 2000; Ching & Foley, 2012; Holland et al., 1998; Peneul & Wertsch, 1995; Wertsch, 1991; Wortham, 2006).

Holland et al. (1998) make an important contribution to the study of identity in their work. It is one that has been echoed by many other researchers, including Boaler and Greeno (2000) in their chapter on identity and agency in mathematics learning. Holland et al. (1998) begin by suggesting that in anthropology there are two dominant schools of thought when thinking about identity and notions of the self. The first school of thought they call "culturalist." In that school of thought culture is something that has a preformed existence and it imposes itself on the individual, shaping their sense of identity. So one has a certain class, racial, gendered, and cultural sense of who one is because of the imposition of culture. The second frame is what they would call "social constructionist." That perspective tends to deconstruct notions of culture and sees forms of identity imposed on and constructed by individuals through discourses of power. They turn to the work of Foucault as one of the dominant theoretical frames for this perspective. We could critique the gloss of these two perspectives as too general and missing the nuances within each perspective, but I don't think that is necessary for my work here.

I would disagree with them on one point. They tend to put psychodynamic theories in the culturalist camp. I would rather tend to see these theories, especially the properly psychoanalytic theories, as a third perspective. They are neither properly culturalist or social constructionist, but rather draw on the ontological fact, as Lacan might say, of humans being born premature but with tremendous cognitive capacity. This basic fact, one might argue, sets the stage for the psychoanalytic unconscious. The psychoanalytic is interesting because each of the perspectives that Holland et al.

lay out, the culturalist and the social constructionist, needs the motor of motivation and desire that tends to come from a more psychodynamic theory. Nevertheless, they do a good job laying out the dilemmas of thinking about identity in anthropology and related fields.

Holland et al. turn to Vygotsky and Bakhtin to provide a new way of thinking about identity. Of course, as we have seen, there are a number of researchers in education and cultural psychology who also have turned to Vygotsky. Holland et al. emphasize Bakhtin's work, which adds another dimension to the thinking about identity. They suggest that the Russians, with their training in Marxist theory, were able to think about the notions of structure, culture, and identity as processes, not static categories. This is similar to our discussion of reification in earlier chapters. By seeing things "in practice" they are able to take on a social constructionist perspective, but one that is dynamic and where there is contestation in the process. And further, there are definite structural and cultural constraints on the individual, but the individual can engage in "semiotic mediation," a term they take from Vygotsky, in order to articulate a path of resistance, avoidance, or otherwise coping with the constraints on individuals (Holland et al., 1998: 38). These practices can be individual or collective, and fairly spontaneous. In this way the cultural worlds are in constant transition; cultures are not static, nor is the individual's identity throughout life. Further, in this way, agency becomes a critical part of thinking about culture as well as about identity.

In order to talk about this notion of individuals creatively mediating the worlds they are part of and the idea of identity in process and culture in process, they use the term "figured worlds." They see figured worlds as:

> a landscape of objectified (materially and perceptibly) expressed meanings, joint activities, and structures of privilege and influence – all partly contingent upon and partly independent of other figured worlds, the interconnections among figured worlds, and the larger society and trans-societal forces.
> (Holland et al., 1998: 60)

In order to think about the structural constraints and the activity of individuals within this concept of figured worlds, they draw heavily on Bourdieu's notions of "habitus" and "field." I talked about Bourdieu in the last chapter on space, since spatial metaphors are so significant in Bourdieu's work. Bourdieu's model of social class that we saw in the last chapter is rooted in the practices of members of the different social class fractions. They then inculcate habits and tastes in the next generation, which he calls habitus. We also discussed Bourdieu's notion of habitus in the last chapter.

Habitus becomes the way that individuals make selections in their lives. The tastes and habits individuals acquire can explain why some kids are good at school and others are not. And even more precisely than that, whether they take a scholarly or a managerial attitude to that work.

Habitus always plays out in some kind of field of social relations. The class system itself is a large field in which individuals interact. But people interact in a lot of fields. The social space of one's workplace, school, profession, community – these are all social fields and often overlapping. In every field there will be an uneven distribution of resources (forms of capital), and the habitus of individual actors will come into play as to how action plays out in that field.

Holland et al. suggest that Bourdieu's work is important for understanding how individuals are caught up in a set of material and symbolic relations and how individuals are predisposed to act in those spaces. Lave and Wenger (1991) make a similar point in their early work on community of practice. They too use Bourdieu to think about the community of practice as a field of relationships and how habitus might shape the individual's actions within that social space.

What is added to this through the notion of figured worlds is the creative, imaginative, unpredictable work that individuals do within those social spaces. They may draw on symbols and materials that are structured in the space, but they do something unpredictable with those resources and they may in fact change the space, help to bring new worlds into being through their actions. In their critique of the structure and agency debate in the social sciences, Emirbayer and Mische (1998) make a similar point. They see too little emphasis on the creative acts of individuals and their ability to do genuinely new things (Charles & Shumar, 2009). Figured worlds is a concept that allows for more agency, but does not dismiss the notion of structure. It attempts to account for how it is possible to bring about new worlds through the artifacts and symbols – as well as constraints – of the current world (Spinosa et al., 1997).

Individualized Communities and Digital Media

The internet and social media have significantly enhanced individuals' abilities to imagine new worlds and creatively work with the resources at hand and the constraints that they experience. Starting in the late 1990s and early 2000s, social network researchers began to understand the powerful potential of the internet for enhancing people's ability to develop and maintain personal networks (Haythornthwaite, 2005; Wellman & Haythornthwaite;

2002; Wellman et al., 1996). In an important article Wellman (2001: 238) talks about the rise of "personal communities." Wellman, like other internet researchers, was aware that the internet changes the nature of social space. His focus was on how geographically bounded communities are no longer the most important way to talk about community. Rather, individuals, through their ability to network online, craft the communities of which they are a part. What Wellman has documented is not just the affinity that individuals have for a common interest, but the connections between members of a network and the way that individuals can craft their own set of relationships and then shape how they access the resources and information of the group.

An important set of concepts from social network analysis is the idea of strong and weak ties: the individuals with whom one has close regular connection and the individuals with whom one's connection is weaker and more irregular. Of course we know that strong ties are important. But Granovetter's (1973) work taught us that weak ties can be important too. Sometimes weak ties are more important, because people who are in the same close community tend to share the same resources and knowledge. New resources and knowledge tend to come from outside one's inner circle. And so a weak tie, when activated, can be an important source for an individual depending on what they are looking for.

Haythornthwaite (2005: 136) further develops this idea with a concept she calls "latent ties." Latent ties are the ties that are created by the introduction of a new medium of communication in a group of people. She further suggests that the new medium may either enhance or disrupt some other weak ties. Haythornthwaite, with this idea, is dealing with a tension in social network theory over the impact of the internet on people's lives. There has been research that has shown that the internet both enhances people's networks but also undermines them. While this is an important issue, it's not my main concern here. My interest is in the two ideas of the rise of personalized communities and the fact that new media create latent ties that have new potential (as well as potential for disruption). These changes in the nature of the social groups that people belong to explain a lot of what we see at the Math Forum and also help us think about the impact of participation in the Math Forum on teachers' identities.

Early on in the Math Forum's history, personalized networks of teachers began to develop. Sometimes these groups were individuals who were part of a project at the Math Forum and who met both face to face and online. The BRAP group, introduced as one of the projects in Chapter 3, is an example of one such group. This group of middle-school teachers who decided to study mathematical discourse in their mathematics classrooms and how that research could be linked to practice was introduced to new media of interaction. While all of them had email accounts and were familiar with web tools, they had not worked so extensively with each other before. This group of teachers became a close and dynamic group. It is clear that the latent ties that were produced by their connection to these new communication tools were activated and they were able not only to work closely with each other but to learn from each other, as they all brought different skills to the table.

In the BRAP project they decided to videotape their classrooms while they were teaching and use these snippets of video as part of their research. This was not only a brave thing to do as individuals, but it added to the rich complexity of forms of communication that were brought into this working group. Finally, there were some interesting shifts in identity. As is common in groups of math teachers, some teachers felt stronger than others in their mathematical ability. Some of the teachers were looked up to as the "math experts" in the group. What the group discovered through working together was that different members of the group brought different skills and expertise to the project. They discovered themselves as experts in this process. But they also discovered each other as resources that they could draw on.

In Chapter 8 we talked about a teacher who felt confident in her abilities to teach algebra and who had taught algebra for a number of years. She was asked to teach geometry and we talked about how she explored an alternative space and community, through the Math Forum, in order to develop the training she needed to feel comfortable teaching in a new area. But another way to think about what has happened here is that she developed a personalized community around her that included a new set of colleagues from which to seek support and advice and a new set of structural resources that contributed to her training. This ability to construct a personalized community also contributed to the teacher's identity work. The teacher was aware that she might be characterized as "not good at geometry" and maybe even more dangerously a "math teacher with limited understandings." We can imagine these characterizations might be gendered too. While these were the potential risks, they were also in the mind of the teacher as a potential way to think about herself. The resources and colleagues that she built online were useful in helping this teacher develop a different identity. It was creative and empowering work that she was able to engage in.

Performances of Self

With the development of social media, this power to craft a personalized community has been enhanced and can be done on the fly. There are now different media through which one can develop professional relationship and share resources. As I suggested at the end of the last chapter, there is now a rhizomatic structure to the communities and social groupings that one finds online. In math education, a number of prominent math teachers have well-known blogs. These blogs are themselves connected to other blogs. Some of the blogs have Twitter accounts. The Mathtwitterblogosphere (http://mathtwitterblogosphere.weebly.com) ties these blogs to other people tweeting about math. And these resources are connected to the Math Forum, NCTM, and other major math education websites.

If we think about these sets of connections from the perspective of one particular individual, we can see something like what network analysts would call an egocentric network. But I am more interested in the communities or groups that the individual connects to and not just the particular individuals. We can see that the community that one individual might craft in this situation is a set of social groups, some larger than others. Or we might also say that the individual belongs to a group of interrelated communities, some larger than others. Certainly, there would be some person-to-person connection in all of this, but not necessarily. A tweet might connect an individual to some resources, a lesson on a blog, a conversation, a group meeting, or any number of other possibilities.

Papacharissi (2012) suggests that in the context of Twitter, the individual engages in a "performance of self." She draws on reflexive modernization theory, which suggests that individuals in the contemporary era have an overwhelming amount of information and that successful individuals are ones that are more reflexive (Beck et al., 1994; Giddens, 1991). Papacharissi (2012: 1990) states: "In late modernity, performances of the self are indicative of the shapes individuals take on as they claim agency and negotiate power within social structures and imaginaries." While Papacharissi cites Goffman (1959) and Butler (1997) when thinking about these performances and what they do, the quote can be linked to our thinking above from Holland et al. (1998). Their notion of a figured world is similar to the thought here. And while the statement has particular implications for Twitter, it is also appropriate for thinking about other social media. Within a set of structural constraints that, thanks to the mobility of the digital world, might be constantly shifting, individuals engage in performances of self that might both be entrapping and/or freeing depending on the creativity of the individual and perhaps the luck of the situation. The context calls for great reflection on the part of individuals, but, of course, the individual cannot always be aware of every implication of every performance of self.

At the Math Forum we have seen not only the spatial shift from a web-centered online community to a more rhizomatic overlapping set of social groups, but also a shift in how math teachers connect with this community and how they engage in a presentation of self. Many leading teachers who are looking for a platform to share their ideas about math and pedagogy have blogs. While the news media suggests the era of blogging is dead or at least dying, blogs are still an active format for a core of math teachers. But the blogs function more as part of web 2.0 than perhaps old-style blogging may have. The blog format persists for math teachers because it's a good format for displaying and discussing math in more detail. It might be about problem solving, visualizing mathematics, or laying out lessons and demonstrations of pedagogy. Many of the math blogs that teachers use look like an Instagram or Twitter feed. And the posting on the blog is linked to these other social media through a posting or tweet that refers to the blog page.

In fact, a group of math teachers have created the Mathtwitterblogosphere, what they refer to as a "global math department." They specifically structure the links, tweets, and blogs into a dynamic set of online interactions through this web-based presentation. They also run a summer face-to-face workshop called Twitter Math Camp, which, of course, has a web and Twitter presence as well. The Mathtwitterblogosphere is a creative way to think about the complex forms of sociality that the interweaving of these technologies can bring. The Math Forum now is a node in this online world. It is certainly a special and large node, but it has become part of something bigger and more distributed online.

In this complex linking of digital tools and the online spaces that are created, math teachers engage in a performance of self, much as Papacharissi suggests. Expert teachers not only can display their knowledge and creative thinking about math pedagogy, but they can do this while choosing to share more informal facts about themselves, their lives, and feelings. Through the build-up of this discourse a sense of a more intimate community is produced. And indeed individuals become closer colleagues and friends through this process. People who have not met before work to meet up in person. The flow of blog posts and tweets also has the effect of creating leaders and experts in the community. Certain teachers are looked up to for their knowledge of math, their ideas about teaching, and so on. A further upside of this online space is that there appears to be more space for women and individuals from diverse cultural backgrounds to take up a discourse of expertise and find voice in this community. And that online expertise could further translate into more respect in their local schools.

There may be others in this space who may not feel that they have the knowledge or creativity to engage in a performance of expertise in this space. But at the same time, they have the potential to perform other roles that will still have the effect of highlighting them as members of this community. Through a process of posting good questions, retweeting, and connecting different individuals, there can be a positive performance where one looks to be a central part of the community, and there may also be the same affective dimensions of sharing greater pieces of one's life. But what is also true is that an individual's sense of intimidation around certain math topics might still remained disguised in this format. In this kind of online performance, one gets to display what one wants and so craft the presentation of self for the community. Of course, according to Goffman (1959), this is always true of individual interaction in any social setting, but the online performance is even more opaque and so allows for greater levels of hiding (Papacharissi, 2012: 1994).

Gender, Race, Math, and Online Interaction

The kind of insecurity around one's math ability, for a teacher or a student, is distributed across social groups. Nevertheless, there are clearly patterns that do have a race, class, and gender dimension to them for teachers and students. In Math Forum workshops there was a tendency for people who taught high school to be more confident of their math ability than teachers who taught in the middle grades. And elementary teachers, not surprisingly, were the least confident. Also in our workshops there was a tendency for this grade level pattern to reflect a gender pattern. More of the high school teachers we worked with were men and more of the elementary teachers were women. And so in workshops you tended to have men being more confident about math and women less confident. There were countertrends to the dominant pattern as well, a strong confident woman high school teacher and a man teaching in elementary school who lacked confidence in mathematics. There was also a social class pattern to our workshop interactions too. Teachers who came from poorer districts, either urban or rural, tended to feel less confident about mathematics and technology. But that was often because they felt their schools did not have the resources to keep up, either for themselves or for their students. The digital divide is also still very real, and these teachers and students often did not have the equipment or the tech support to take advantage of online resources and interactions.

Many of the teachers who felt they did not have a strong math background were able to take up other roles, especially in online work that was part of a larger project. These teachers often performed leadership or social roles, organizing groups, maintaining resources, and making connections between people. This gave them a presence in the online community but did not put too much pressure on them mathematically. This could be seen as a form of social hiding, but I think for many of the teachers it was productive for them to be part of the community and be able to have voice and increase their skills and sense of confidence. The teachers with less confidence also greatly benefited from the Noticing & Wondering framework. As I discussed earlier, one of the powers of the N&W framework is that it is an easy scaffold for beginning to have something to say about problems and problem solving. It not only empowers a person to begin to have valuable things to say about a problem, but also leads them to making connections that they may not have made if they had not tried this scaffold. N&W also gave these teachers ideas about how to work with their students around a math problem and gave them confidence about starting a discussion around math, and not just following a procedure.

As I said at the beginning of the book my emphasis in this monograph has been on the teachers who are part of the Math Forum. They are the most regular and persistent part of the community, and many of the Math Forum workshops and projects worked with teachers. But of course the Math Forum works with students too. For example, the Problems of the Week were problems that primarily students solved. The Virtual Math Teams project discussed in Chapter 3 is one of the Math Forum projects that is exclusively focused on students and their problem solving. Further, as a synchronous problem-solving environment the director of the Math Forum suggests that one has a real-time view of how students do problem solving, which can be replayed since the synchronous sessions are recorded. It gives the mentor or teacher an opportunity to look multiple times at the process of students doing work together and query that work.

When the Math Forum was at Drexel University, the Philadelphia school district was taken over by the state. Part of the state's plan was for various stakeholders to work with or manage different groups of schools in the district. One of those groups was the Philadelphia Temple Partnership, and the Math Forum worked directly with this group. This work was more face-to-face work than online, as Math Forum staff members worked in different classrooms with students and teachers. The classrooms were made up of students, mostly of color, from poor neighborhoods in North Phila-delphia. In one particular eighth grade class, the Math Forum was doing N&W with a group of students. They had been doing this work for a number of days when the director of the Math Forum visited the class and looked at the work different groups of students were doing. He was impressed with the way they had leveraged the N&W model to advance their own mathematical work. The groups had made a lot of progress on their own. He mentioned to one group that he was very impressed with the progress they had made and started to ask them about it. They responded to him firmly to not interrupt them, that N&W was their process and they were not yet finished with probing their own insights about the math.

What further impressed the director was that the students not only had made some excellent progress thinking about mathematics on their own, but felt strongly about owning the process. N&W was their process and they wanted to continue to work through that process before they shared their insights about the math they were doing. In this instance, with a group of kids who in terms of statistics are thought of as poor performers in math, the kids felt agentic in their math work and the N&W scaffold worked exactly as the Math Forum staff hoped it would. It became a ramp for them to begin to have conversations about what they saw and what they could think about, but that then led to a greater comfort with the deeper conversation about math. It also led to a comfort with owning the process of talking about math, which does happen for kids in group work, but this was a particularly powerful instance of that ownership. Echoing this story and others at the Math Forum, Boaler and Greeno (2000: 172), who also use the notion of "figured worlds" to frame their research, suggest, like other social learning theorists, that learning mathematics is linked to the doing of math and talking about mathematics.

They further use the notion of what they call "ecologies of participa-tion" and think about how the context positions students to be active participants and agents in their own mathematical practice, discourse, and thinking. It would be interesting if this group of students in the Phila-delphia school district had the teachers and their students had the resources to continue to use the N&W framework online as well as take advantage of the online resources. The Math Forum model could really have an impact on inner city classrooms.

The focus of Math Forum as a research project and an online organiza-tion has never been to specifically target race and class inequality. Never-theless a number of the staff members have training in social justice perspectives and have worked with organizations that seek to promote a more equitable education. They have worked with a number of inner city schools, including the early Urban Systemic Initiative schools discussed in Chapter 3, some in-depth work with the Union City school district in New Jersey, and then this more recent work with the Philadelphia Temple Partnership. In all of these instances, there was a face-to-face component and an online component to the work. The Math Forum model of focusing on relationships between individuals and developing a mathematical conversation with students and teachers has had a powerful impact not just on the kind of mathematical work the teachers and students could do together but also on the way they thought about themselves as being empowered to do this work. Of course, in all school districts there are real pressures to keep from doing this unalienated kind of work, and those pressures are tied up with limitations of time and resources. People feel they do not have the time to encourage students to be mathematical thinkers but rather need to meet certain numerical targets on standardized exams. And that pressure means that work needs to move quickly, and there is a tendency toward a focus on procedures rather than understanding. This is a problem that is well known in the math education community.

The Entrepreneurial Self and Neoliberalism

One of the things we have clearly seen in this work with the Math Forum is that the internet and digital technologies, when used to enhance opportunities to talk and interact with others, and when the focus is on a community of practice and how to make optimal use of new technologies for learning, are tremendous opportunities for individuals. The technologies can lend themselves to the enhancement of selves because scaffolds can be personalized; there are multiple channels for conversation and different kinds of discourses. The medium lends itself to the Math Forum processes of double reflection and to balance reifications with interactions.[1]

This potential of online learning and digital technologies plays very well into current research in educational psychology where concepts such as resilience, motivation, and creativity characterize much of the current thinking. There is a real effort to understand what psychologists might call self-efficacy among students and why some students can work independently, persist in their work, and make progress, while other students seem to have a great deal of trouble doing this. As I have already discussed elsewhere, online learning can be structured to be an ideal scaffold for students, where students have enough support to work independently and that support can be removed when appropriate or shifted to meet the needs of particular students. In my work with teachers, I would say the support scaffolds work in the same way, where teachers did not even fully realize they had transformed into someone who felt more confident about their ability to work with math and about their skills with technology (Renninger & Shumar, 2002, 2004). Further, we have seen that this sense of self that individuals have when they are empowered to work on their own and make progress give people a real sense of agency in the ability to do math work and to think of themselves as good at math. We have seen this in both teachers and students.

In the internet world the emphasis on individual agency and the potential of the internet to help people craft their own communities or social groups sometimes can lead individuals to thinking that there are no structural forces operating. But it is clear from a broad range of research looking into patterns of online interaction that structural forces are still very much at work. While the broad contours of the digital divide may be closing, with a much higher percentage of people going online and a high percentage of people accessing the internet through mobile technology, poor schools still have a tough time taking advantage of the online resources available to them. And while the internet can reduce social cues and status markers, as we have seen in the Math Forum research, we know have a large body of research that shows there are still clear patterns of gendered research online. We can infer that other social cues are also readable online as well. Finally, Bourdieu's notions of cultural and linguistic capital help us understand that it is much more likely that individuals from higher socioeconomic backgrounds will be better able to leverage the resources online and the opportunities for online educational conversations.

This tension between what we might think of as the neoliberal tendencies in discourse around the internet and digital media, and the genuine protean possibilities for individual growth and creativity, is an important issue for thinking about the future of the internet and education (Lifton 1993). Many universities and schools these days tend to see education as a commodity. And if education is a commodity that can be sold to learners in different contexts, then the digital is seen as an efficient medium of exchange for those commodities. The problem with this view, as we have shown, is that learning and knowledge building are dynamic social processes and not a matter of consuming commodities. In order to really think about the kinds of value that digital technologies bring to learning and interaction, we need to move way beyond the ways in which we tend to see online learning and online courses and see them less as analogues of face-to-face courses and more in the way the Math Forum sees the digital.

The Math Forum's prescription, if you will, for supporting creative and empowered individuals has been to minimize the effect of the more negative structural forces. This was often done by its creative response to structural inequality and finding ways to work around conditions where there are limitations of hardware, software, or technical support. It also always tended to reduce the visibility of status differences both online and in face-to-face workshops. At the same time, it used the structural support of digital technologies to encourage everyone to notice and wonder things about a problem or a scenario and encourage a robust conversation around those noticings. These processes worked well for all of the teachers and students they worked with regardless of their backgrounds. But it is true that larger structural inequities were ones that the Math Forum did not directly have the power to take on.

Summary

In Chapter 4 we talked about the contradictions of the digital economy at this moment in our global economic system and how the Math Forum has been caught in that dialectic. The tensions the Math Forum has felt between the utopian potential of the internet and the rush to reify digital commodities is paralleled by the experience of individuals in the Math Forum's online educational community.

On the one hand, the power to craft more personalized communities and to create overlapping groups of communities that share knowledge and resources is a powerful contribution to the creative potential of individuals. Further, the different ways that individuals can engage in a "presentation" or "performance of self" also makes room for the creative and empowered building of one's own knowledge base in mathematics and math education.

At the same time, there are pressures on individuals in this context to commodify themselves, as Ilana Gershon (2011) would say, to see themselves as a business. An implication of Gershon's idea, that we are all hailed as neoliberal agents, is a denial of the social. There are no structural or cultural differences between people, there are only individuals who are competing with each other in a marketplace to be the best math student, the best math teacher, the best employee. And the internet and digital technologies can facilitate that exchange.

The tensions between the commodified self and the empowered self are often difficult to see, as we tend to use the same language. And as the factors that divide those who have from those who have not become more subtle in an online context, it's easy for people to dismiss structural differences. This is an issue that the Math Forum will have to find ways to address in its future work.

Note

[1] Not that this is the dominant way that the educational community thinks about online learning and digital technologies. There is still a strong tendency to think that the resource (lecture, lesson, problem) is the most important piece, and so the reified object becomes fetishized and the interaction between individuals is seen as secondary. This is what I mean by the reversal of the balance between reification and interaction. And this mode of seeing the proliferation of resources as the most important thing about digital technologies for learning has led to fantasies of online education replacing face-to-face education. The excitement over MOOCs recently is an example of such thinking.

10   Conclusion

In this chapter I will pick up where I left off in Chapter 9. We will then go on to review the potential of online learning and think a bit about the future of the Math Forum. The Math Forum has been an amazingly durable institution. But as a small organization it has been reduced by some of the battles it has had to endure. For its future to be successful, it will need to find ways to build back up as an organization and move forward with its work.

Contradictions in the Internet Economy

The Math Forum is an entrepreneurial organization. It is creative and flexible. Its members have been very innovative in the work they do, and also in their efforts to be successful as a venture. In other work I have been doing, I have been thinking, along with colleagues in Europe, about the meaning of the terms "entrepreneurial" and "entrepreneurship." Ron Barnett (2013) suggests that "entrepreneurial" is a captive signifier in that the architects of the global economy want entrepreneurial to mean neo-liberal agents in Gershon's (2011) sense, where entrepreneurs are creating ventures that are primarily focused on capital accumulation. The critics of this vision, by and large, accept this definition of entrepreneurial and focus on how this vision undermines the well-being of large numbers of people in order to concentrate capital accumulation for the few. For Barnett (2013) this is an ideological closure around the concept; entrepreneurial is only connected to neoliberal capitalism.

As discussed in Chapter 4, a number of researchers have looked at how the development of the internet, coming closely on the stagnation of the postwar Fordist system, became central to a vision where creativity and entrepreneurship could be enhanced by digital technologies while contributing to the accumulation of profit (Neff, 2012; Schiller, 2000).

In this view, the old contract that employers had with employees, where job security and benefits were part of the deal, could be severed in the name of greater creative freedom and the potential to be wildly successful as a creative individual (Neff, 2012). While this new economy did create winners, especially in its earlier phases, there were also many losers. And now, as global inequality has increased dramatically in the last thirty to forty years, the losers far outnumber the winners (Piketty, 2014).

While teachers and students are often not directly hailed as neoliberal subjects, they are encouraged toward a neoliberal view of value and entrepreneurship through the audit and accountability mechanisms that measure individuals and their learning by scores on standardized exams. These views are also affecting universities worldwide as research productivity, teaching, and learning are measured by the same kinds of market-based standards. These kinds of neoliberal pressures are destructive to education because they incorrectly measure the wrong kinds of things when trying to think about learning and knowledge production.

But colleagues in Scandinavia and other parts of Europe have a much broader definition of entrepreneurial and entrepreneurship. For them entrepreneurship is about creating value in a much broader sense (Shumar & Robinson, forthcoming). This broader sense of entrepreneurship is an idea that is shared among individuals in the craft and DIY economies, as well as among individuals who might think of themselves as social entrepreneurs. In this model, value can be things that contribute more broadly to the well-being of a community of people. In fact, we could think of the accumulation of profit, from this perspective, as creating one kind of value while undermining other values such as the quality of the local environment or the quality of the life of a group of people who live in the area. Ultimately, wealth production might be a net loss in terms of value given that profit has such a narrow range of benefit.

It is in this broader sense of entrepreneurial that I see the Math Forum as an entrepreneurial organization. They have struggled with the pressures to pay for the kinds of work they do, and they have struggled with the neoliberal imagination of the internet and how, as an internet organization, they would just naturally find their way to profit. But at the same time, they have found a way to survive and they have pointed the way toward how the internet and digital tools can support more complex learning communities that have more ways to work together and produce knowledge. In Spinosa and colleagues' (1997) sense, they have disclosed a new world.

As universities have been pressured to see themselves even more as a business, it's been more difficult for the Math Forum to fit within the university structure. From the beginning, it was always Swarthmore College's intention to "spin them off," although that was seen more as a "type of university" issue. Since the college was a small liberal arts organization, they did not see a research project as fitting there in the long term. Nevertheless, everyone felt at the time that internet organizations such as the Math Forum should be self-sustainable. At Drexel University the fit was always uneasy. I had always thought of the Math Forum as a research incubator, itself a kind of neoliberal idea, but it was my way of imagining their place in the university. But as Drexel, like so many universities, adopted Responsibility Centered Management (RCM), it became even harder to imagine how it fit there because the Math Forum did not have student customers and they did not directly support units that did have student customers.

As the Math Forum moves from Drexel University to the National Council of Teachers of Mathematics (NCTM), it seems like the perfect home for them. An organization long dedicated to reform pedagogy in mathematics and supporting math teachers, the fit appears to be near perfect. As a professional organization NCTM perhaps can be shielded from some of the neoliberal pressures such as RCM. But of course, they must deal with the pressures to assess the work of teachers and students, and so audit and accountability are pressures that will continue to affect them and the Math Forum.

Potential of Online Learning

In this book I have tried to talk about some of the key contributions the Math Forum has made to mathematics education and the potential of the internet and digital technologies. Some of the core ideas about mathematics education being rooted in practice, discourse being central to the learning process, Noticing & Wondering being a valuable scaffold for entering into mathematical dialogue for teachers and students are all echoed by other scholars in math education, especially those from a social learning perspective.

Likewise, what I have been calling the Math Forum dialectic, with its emphasis on practice, discourse, and thinking, could be seen as similar to, and even informed by, ideas that other scholars have had (Boaler, 2000; Cobb et al., 2000; Sfard, 2008). What the Math Forum uniquely brings is its unwavering focus on the right balance between reification and interaction (Wenger, 1998). The conversation and interaction is where knowledge is built. Resources and tools can support that interaction and attempt to capture stages of the knowledge-building process, but the interaction and what people are thinking is the primary focus. Further online tools change the social space in which these interactions take place and change the kind of conversations or discursive interactions that people can have with each other. To me these are key things that so many other online learning efforts fail to understand. The dialectic is not just that each leg of the triangle leads back to the other legs, but also that each produces and is a product of the other; for example, practice produces discourse, and discourse leads back to and supports practice. Reifications that are produced through the process of moving from practice to discourse to thinking support knowledge moving up and are deconstructed as people continue to evolve their thinking.

The process of Noticing & Wondering and what I have been calling "double reflection" are key aspects to the Math Forum process. As I have said, I think they are different from teacher professional noticing, and because the Math Forum had the freedom to develop this practice outside a classroom context, they bring a novel approach that, at least in part, could be integrated into a classroom context. N&W begins as an internal process. It starts with what the individual notices and wonders about the problem context. And it is a highly reflexive process: What do I see here? What questions could I ask? What directions could I take this process toward? and What other knowledge do I have to bring to this situation? N&W not only is directed at one's own problem solving but gets directed at the other (Ray-Riek, 2013). This is always consciousness about some objects; consciousness about the objects (symbols, mathematical ideas, etc.) of inquiry as well as the objects themselves are each formed in the process of the reflection. Neither is preexisting; both are formed in context. In that way, this is a Hegelian process. And while we can think about it in complex philosophical terms, as I have also said, it's a practical, down-to-earth process. The N&W scaffold allows individuals to engage in the process of double reflection from a simple starting point. As the director of the Math Forum has said to me many times, "This is a normal human process. It is the way people explore and come to understand things."

Through a focus on user-generated content the Math Forum has also developed a rich set of resources. Workshops at the Math Forum were not only hybrid, in that they had face-to-face and online components, and long lasting, in that projects often continued for weeks and sometimes for the individuals became much longer projects, but they tended to produce curricula and other materials that could be part of the site's resource base. We talked earlier about how Ask Dr. Math and the PoWs developed a rich set of resources through the repurposing of interactions in those spaces. Further, the Math Forum collected materials from other organizations that wanted their materials linked to the Math Forum or were looking for a site to host resources. So the Math Tools site has many donated apps that are part of the collection. These activities are very much the way of the internet, and many sites reuse interactions, make up FAQs from earlier queries, or remix materials to create new digital objects. But in math education the Math Forum was one of the first to do this and is still one of the premier educational collections.

Spaces for Transformation

The Math Forum has really been a model of a transformative online space. I think the most critical piece of this is that the staff has kept their eyes on how to improve opportunities for people to do mathematical activities and talk together. Further, their assumption, that math is part of everyone's lives, leads them toward an attitude toward people that is about getting to know the person. In this way, I find them to be an exemplary community for an anthropological analysis of the potential of the digital for learning. Their attitude toward the digital has helped them be creative about the kinds of online spaces they inhabit and what they do in those spaces. When the team works together there is often a working web page or wiki that is part of the meeting process. Even when they are all together in one room with each other, they are in a virtual space too. NSDL was an important idea and an important part of shaping the Math Forum's thinking about space. The Math Forum may no longer think of themselves as an inter-active digital library. But that moment helped frame the importance of good resources and the importance of virtual spaces to interact with those resources. It allowed them to connect their ideas about interactivity and how resources should be integrated into interactivity into a larger national agenda.

As I said, from the beginning, the Math Forum was not trying to create an online community. But rather it had an unalienated view of human interaction and human potential. While the staff would say that the focus on problem-solving, communication, and creating valuable resources is the thing that created the community, it is also clear that they very much value the community. Over time, the Math Forum has become sensitive to how different tools and technologies allow for creating different kinds of social spaces. And different spaces have different kinds of opportunities for learning as well as different kinds of opportunities for self-development.

Much of online learning in higher education is behind the Math Forum and its thinking about learning in online spaces. The field could learn a lot from paying attention to Math Forum practices and adopting some of those practices. A number of the efforts to create MOOCs, for instance, have fallen into the trap of putting the objects, the reification – for example, lectures, models, syllabi – before the human interaction. This fetishizing the object does not encourage people to engage in the practice of producing meaning together. And so MOOCs are not about knowledge production; rather, they are about the transmission of information. And as we have seen the transmission of information has not worked very well. This is not to say there are not efforts to create more dynamic learning communities in MOOCs. But the excitement has been more on the stuff, and less on people interacting together.

The Future of the Math Forum

At the writing of this book, the Math Forum has just finished its transition to NCTM. I am very excited for this moment for them. They began before the web, and now almost twenty-five years later, they are about to embark on a new chapter. As I have tried to show throughout this book, the Math Forum has achieved some incredible things. There are many things that we can learn about creating online education communities from them. And there are many things we can learn from them about how technology can aid a human-centered learning, a process of learning that involves people working and talking together, producing meaning and knowledge intersubjectively.

But the Math Forum also experienced limits to its horizon. Unlike Facebook or Amazon, its members did not have the resources to see where their vision would take them before they focused on making money. Making money, enough money to pay the bills, became a necessary focus of the Math Forum for most of its time. And this focus limited the horizons of what it could do with online learning resources and technology in support of math education. I am sure, now that it is at NCTM, that the challenges of paying the bills will still be part of its reality and a pressure on its activities. It is my hope that because there is a greater alignment with the Math Forum's goals and NCTM's goals there will be more opportunities to move the practice of supporting math education further. It's an exciting time for the Math Forum, and I look forward to the next chapter in its story.

References

Anderson, B. R. O. G. (1991). *Imagined Communities: Reflections on the Origin and Spread of Nationalism.* Revised and extended ed. New York: Verso.

Appadurai, A. (1990). Disjuncture and difference in the global cultural economy. *Theory, Culture & Society* 7:295–310.

Bahktin, M. M. (1986). *Speech Genres and Other Late Essays.* Austin: University of Texas Press.

Bakhtin, M. M., & M. Holquist, (1981). *The Dialogic Imagination: Four Essays,* Austin: University of Texas Press.

Ball, D. L., & Bass, H. (2000). Interweaving content and pedagogy in teaching and learning to teach: Knowing and using mathematics. In Boaler, J. (Ed.), *Multiple Perspectives on Mathematics Teaching and Learning* (pp. 83–104). Westport, CT: Ablex Publishing.

Ball, D., Hill, H., & Bass, H. (2005). Knowing mathematics for teaching: Who knows mathematics well enough to teach third grade, and how can we decide? *American Educator* 29(3):14–22, 43–46.

Barab, S. A., Thomas, M. K., Dodge, T., Squire, K., & Newell, M. (2004). Critical design ethnography: Designing for change. *Anthropology & Education Quarterly* 35(2):254–268.

Barab, S., Grey, J., & Kling, R. (Eds.) (2004). *Designing for Virtual Communities in the Service of Learning.* New York: Cambridge University Press.

Barnett, R. (2013). *Imagining the University.* Milton Park, Abingdon, Oxon, and New York: Routledge.

Bauman, Z. (2000). *Liquid Modernity.* Cambridge: Polity Press; Malden, MA: Blackwell.

Beck, U., Giddens, A., & Lash, S. (1994). *Reflexive Modernization: Politics, Tradition and Aesthetics in the Modern Social Order.* Stanford: Stanford University Press.

Bell, P. (2012). Understanding how and why people learn across settings as an educational equity strategy, in Bevan, B., Bell, P., Stevens, R., & Razfar, A. (Eds.), *Learning about Out of School Time (LOST) Learning Opportunities* (pp. 224–241). London: Springer.

Bhabha, H. K. (1994). *The Location of Culture.* London and New York: Routledge.

Bluestone, B., & Harrison, B. (1982) *The Deindustrialization of America: Plant Closings, Community Abandonment, and the Dismantling of Basic Industry*. New York: Basic Books.

Boaler, J., & Greeno, J. (2000). Identity, agency and knowing in mathematics worlds. In Boaler, J. (Ed.), *Multiple Perspectives on Mathematics Teaching and Learning* (pp. 171–200). Westport, CT: Ablex Publishing.

Boellstorff, T. (2008). *Coming of Age in Second Life: An Anthropologist Explores the Virtually Human*. Princeton: Princeton University Press.

    (2012). Rethinking digital anthropology. In Horst, H. A., & Miller, D. (Eds.), *Digital Anthropology*. London and New York: Berg Publishers.

Bourdieu, P. (1990). *The Logic of Practice*. Cambridge: Polity Press.

Bourdieu, P., & Wacquant, L. J. D. (1992). *An Invitation to Reflexive Sociology*. Chicago: University of Chicago Press.

Brown, J. S., Collins, A., & Duguid, P. (1989). Situated cognition and the culture of learning. *Educational Researcher* 18(1):32–42.

Bruner, J. (1966). *Toward a Theory of Instruction*. Cambridge, MA: Harvard University Press.

    (1996). *The Culture of Education*. Cambridge, MA: Harvard University Press.

Butler, J. (1997). *The Psychic Life of Power: Theories of Subjection*. Stanford: Stanford University Press.

Castells, M. (1989). *The Informational City: Economic Restructuring and Urban Development*. Oxford: Blackwell Publishers.

    (2001). *The Internet Galaxy*. Oxford: Oxford University Press.

Chaiklin, S., & Lave, J. (Eds.) (1996). *Understanding Practice: Perspectives on Activity and Context*. Cambridge: Cambridge University Press.

Charles, E. S., & Shumar, W. (2009). Student and team agency in VMT. In Stahl, G. (Ed.), *Studying Virtual Math Teams*, Computer-Supported Collaborative Learning Series, Vol. 11. New York: Springer.

Ching, C. C., & Foley, B. J. (2012). Introduction: Connecting conversations about learning, identity. In Ching, C. C., & Foley, B. J. (Eds.), *Constructing the Self in a Digital World*. Cambridge: Cambridge University Press.

Clifford, J. (1988). *The Predicament of Culture: Twentieth-Century Ethnography, Literature, and Art*. Cambridge, MA: Harvard University Press.

Clifford, J., & Marcus, G. (1986). *Writing Culture: The Poetics and Politics of Ethnography*. Berkley: University of California Press.

Cobb, P., Gresalfi, M., & Hodge, L. L. (2009). An interpretive scheme for analyzing the identities that students develop in mathematics classrooms. *Journal for Research in Mathematics Education* 40(1):40–68.

Cobb, P., Yackel, E., & McClain, K. (2000). *Symbolizing and Communicating in Mathematics Classrooms: Perspectives on Discourse, Tools and Instructional Design*. Mahwah, NJ: Lawrence Erlbaum Associates Publishers.

Cole, M. (1998). *Cultural Psychology: A Once and Future Discipline*. Cambridge, MA: Belknap Press, Harvard University.

Coleman, G. (2010). Ethnographic approaches to digital media. *Annual Review Anthropology* 39:487–505.

D'Andrade, R. G. (1995). *The Development of Cognitive Anthropology*. Cambridge: Cambridge University Press.

Davies, B., & Harré, R. (1990). Positioning: The discursive production of selves. *Journal for the Theory of Social Behavior* 20:43–63.

Deleuze, G., & Guattari, F. (1987). *A Thousand Plateaus: Capitalism and Schizophrenia*. Translation and foreword by Brian Massumi. Minneapolis: University of Minnesota Press.

Dewey, J. (1938). *Experience and Education*. New York: Macmillan.

Emirbayer, M., & Mische, A. (1998). What is agency? *American Journal of Sociology* 103(4):962–1023.

Fabian, J. (1983). *Time and the Other: How Anthropology Makes Its Object*. New York: Columbia University Press.

Falk, J., & Drayton, B. (Eds.) (2009). *Creating and Sustaining Online Professional Learning Communities*. New York: Teacher College Press.

Faubion, J. (2009). The ethics of fieldwork as an ethics of connectivity, or the good anthropologist (isn't what she used to be). In Faubion, J., & Marcus, G. (Eds.), *Fieldwork Is Not What It Used to Be: Learning Anthropology's Method in a Time of Transition*. Ithaca, NY: Cornell University.

Fetter, A. (2008). Using the PoWs: Getting started: How to start problem solving in your classroom. Available at http://mathforum.org/pow/teacher/PoWsGetting Started.pdf

Florida, R. (2005). *Cities and the Creative Class*. New York and London: Routledge.

Gee, J. P. (2004). *Situated Language and Learning: A Critique of Traditional Schooling*. New York: Routledge.

(2005). Semiotic social spaces and affinity spaces: From the age of mythology to today's schools. In Barton, D., & Tusting, K. (Eds.), *Beyond Communities of Practice: Language, Power and Social Context* (pp. 214–232). Cambridge: Cambridge University Press.

(2007). *What Video Games Have to Teach Us about Learning*, 2nd ed. New York: St. Martin's Press.

Gershon, I. (2011). Neoliberal agency. *Current Anthropology* 52(4):537–555.

Gibson, W. (1984). *Neuromancer*. New York: Ace Books.

Giddens, A. (1991). *Modernity and Self-Identity: Self and Society in the Late Modern Age*. Palo Alto, CA: Stanford University Press.

Goffman, E. (1959). *The Presentation of Self in Everyday Life*. New York: Doubleday.

Goodwin, C. (1994). Professional vision. *American Anthropologist*, 96(3):606–633.

Gottdiener, M. (1993). A Marx for our time: Henri Lefebvre and the production of space. *Sociological Theory* 11(1):129–134.

Graham, M., & Dutton, W. H. (Eds.) (2014). *Society and the Internet: How Networks of Information and Communication Are Changing Our Lives*. Oxford: Oxford University Press.

Granovetter, M. S. (1973). The strength of weak ties. *American Journal of Sociology*, 78:1360–1380.

Gupta, A., & Ferguson, J. (1997). *Anthropological Locations: Boundaries and Grounds of a Field Science*. Berkeley: University of California Press.

Habermas, J. (1984). *The Theory of Communicative Action: Reason and the Rationalization of Society*, Vol. 1. Trans. Thomas McCarthy. Boston: Beacon Press.

(1989). *The Structural Transformation of the Public Sphere: An Inquiry into a Category of Bourgeois Society*. Cambridge, MA: Polity Press.

Hardt, M., & Negri, A. (2000). *Empire*. Cambridge, MA, and London: Harvard University Press.

Harré, R., & Van Langenhove, L. (1991). Varieties of positioning. *Journal for the Theory of Social Behavior* 21(4):393–407.

Harvey, D. (1990). *The Condition of Postmodernity: An Enquiry into the Origins of Cultural Change*. Oxford, UK: Blackwell Publishers.

(2000). *Spaces of Hope*. Berkeley: University of California Press.

(2006). *Spaces of Global Capitalism: Toward a Theory of Uneven Geographical Development*. London, New York: Verso.

Haythornthwaite, C. (2005). Social networks and internet connectivity effects. *Information, Community & Society* 8(2):125–147.

Heidegger, M. (1962, original 1927). *Being and Time*. New York: Harper and Row.

Hiebert, J. (Ed.) (1986). *Conceptual and Procedural Knowledge: The Case of Mathematics*. New York and London: Routledge.

Hill, H. (2010). The nature and predictors of elementary teachers' mathematical knowledge for teaching. *Journal for Research in Mathematics Education* 41(5):513–545.

Hill, H. C., & Ball, D. L. (2004). Learning mathematics for teaching: Results from California's Mathematics Professional Development Institutes. *Journal for Research in Mathematics Education* 35(5):330–351.

Hill, H. C., Rowan, B., & Ball, D. L. (2005). Effects of teachers' mathematical knowledge for teaching on student achievement. *American Educational Research Journal* 42(5):371–406.

Hine, C. (2000). *Virtual Ethnography*. London, Thousand Oaks, New Delhi: Sage.

(2015). *Ethnography for the Internet: Embedded, Embodied and Everyday*. London: Bloomsbury.

Hogan, M., & Alejandre, S. (2010). Problem solving – it has to begin with noticing and wondering. *CMC ComMuniCator, Journal of the California Mathematics Council* 35(2):31–33.

Holland, D., Lachicotte, W. Jr., Skinner, D., & Cain, C. (1998). *Identity and Agency in Cultural Worlds*. Cambridge, MA: Harvard University Press.

Jackson, J. L. (2012). Ethnography is, ethnography ain't. *Cultural Anthropology* 27(3):480–497.

Jacobs, V., Lamb, L., & Philipp, R. (2010). Professional noticing of children's mathematical thinking. *Journal for Research in Mathematics Education*, 41(2): 169–202.

Jones, S. (Ed.) (1998). *Cybersociety 2.0: Revisiting Computer-Mediated Community and Technology*. Thousand Oaks: SAGE Publications.

Kirschner, D., & Whitson, J. A. (Eds.) (1997). *Situated Cognition: Social, Semiotic and Psychological Perspectives*. Mahwah, NJ: Lawrence Erlbaum Associates.

Kollock, P. (2002). The economies of online cooperation: Gifts and public goods in cyberspace. In Kollock, P. & Smith M.A. (Eds.), *Communities in Cyberspace*. New York: Routledge.

Kollock, P., & Smith M. A. (Eds.) (2002). *Communities in Cyberspace*. New York: Routledge.

Lakoff, G., & Johnson, M. (2003). *Metaphors We Live By*, 2nd ed. Chicago: University of Chicago Press.

Lareau, A. (2011). *Unequal Childhoods: Class, Race, and Family Life*, 2nd ed. with an update a decade later. Berkeley: University of California Press.

Lave, J. (1997). The culture of acquisition and the practice of understanding. In Kirshner, D., & Whitson, J. A. (Eds.), *Situated Cognition: Social, Semiotic, and Psychological Perspectives* (pp. 17–36). Mahwah, NJ: Lawrence Erlbaum.

Lave, J., & Wenger, E. (1991). *Situated Learning: Legitimate Peripheral Participation*. Cambridge: Cambridge University Press.

Lefebvre, H. (1991). *The Production of Space*. Oxford: Blackwell Publishers.

Lifton, R. J. (1993). *The Protean Self: Human Resilience in an Age of Fragmentation*. New York: Basic Books.

Lowenthal, D. (1999). *The Past Is a Foreign Country*. Cambridge: Cambridge University Press.

Lukács, G. (1972). *History and Class Consciousness: Studies in Marxist Dialectics*. Cambridge, MA: MIT Press.

MacLeod, J. (2008). *Ain't No Makin' It: Aspirations and Attainment in a Low-Income Neighborhood*, 3rd ed. Boulder: Westview Press.

Marcus, G. (2010). Notes from within a laboratory for the reinvention of anthropological method. In Melhuus, M., Mitchell, J. P., & Wulff, H. (Eds.), *Ethnographic Practice in the Present*. New York: Berghahn Books.

(2012). Multi-sited ethnography: Five or six things I know about it now. In Coleman, S., & von Hellermann, P. (Eds.), *Multi-Sited Ethnography: Problems and Possibilities in the Translocation of Research Methods*. New York: Routledge.

Marcus, G., & Fischer, M. (1986). *Anthropology as Cultural Critique: An Experimental Moment in the Human Sciences*. Chicago: University of Chicago Press.

Markoff, J. (2006). *What the Dormouse Said: How the Sixties Counterculture Shaped the Personal Computer Industry*. San Francisco: Penguin Group USA.

Mathews, J. (2005). 10 myths (maybe) about learning math. *Washington Post*, May 31, 2005. www.washingtonpost.com/archive/business/technology/2005/05/31/10-myths-maybe-about-learning-math/0b7b25ca-baf8-4982-8e9e-fa644cc369b2/.

Miller, D. (1987). *Material Culture and Mass Consumption*. Oxford: Blackwell.

Miller, D., & Horst, H. A. (2012). The digital and the human: A prospectus for digital anthropology. In Horst, H. A. & Miller, D. (Eds.), *Digital Anthropology*. London and New York: Berg Publishers.

Miller, D., & Slater, D. (2000). *The Internet: An Ethnographic Approach*. Oxford and New York: Berg Publishers.

Neff, G. (2012). *Venture Labor: Work and the Burden of Risk in Innovative Industries*. Cambridge, MA: MIT Press.

Noble, D. F. (2002). *Digital Diploma Mills: The Automation of Higher Education*. New York: Monthly Review Press.

Oldenburg, R. (1989). *The Great Good Place: Cafes, Coffee Shops, Bookstores, Bars, Hair Salons, and Other Hangouts at the Heart of a Community*. New York: Marlowe and Company.

Ong, A. (2006). *Neoliberalism as Exception Mutations in Citizenship and Sovereignty*. Durham, NC: Duke University Press.

Papacharissi, Z. (2012). Without you, I'm nothing: Performances of the self on Twitter. *International Journal of Communication*, 6:1989–2006.

Pea, R. D. (2004). The social and technological dimensions of scaffolding and related theoretical concepts for learning, education, and human activity. *Journal of the Learning Sciences* 13(3):423–451.

Peirce, C. S. (1982). *The Writings of Charles S. Peirce: A Chronological Edition*, Vol. 2. Ed. Peirce Edition Project. Bloomington: Indiana University Press.

Peirce, C. S., Hartshorne, C., Weiss P., & Burks, A. W. (1931). *Collected Papers of Charles Sanders Peirce*. Ed. Charles Hartshorne and Paul Weiss. Cambridge, MA: Harvard University Press.

Penuel, W. R., & Wertsch, J. V. (1995). Vygotsky and identity formation: A sociocultural approach. *Educational Psychologist* 30(2):83–92.

Peters, M. A., Marginson, S., & Murphy, P. (2009). *Creativity and the Global Knowledge Economy*. New York: Peter Lang Publishers.

Piketty, T. (2014). *Capital in the Twenty-first Century*. Cambridge, MA: Belknap Press, Harvard University.

Poster, M. (2001). *What's the Matter with the Internet?* Minneapolis: University of Minnesota Press.

Putnam, R. D. (2001). *Bowling Alone: The Collapse and Revival of American Community*. New York: Simon and Schuster.

Ray-Riek, M. (2013). *Powerful Problem Solving: Activities for Sense Making with the Mathematical Practices*. Portsmouth, NH: Heinemann.

Regis, T., Silverman, J., Clay, E., Shumar, W., & Madison, N. (2009). The Math Forum's online professional development model to identify mathematics leaders. In I. Gibson et al. (Eds.), *Proceedings of Society for Information Technology and Teacher Education International Conference 2009* (pp. 2943–2946). Chesapeake, VA: AACE.

Renninger, K. A., & Shumar, W. (Eds.) (2002). *Building Virtual Communities*. New York: Cambridge University Press.

  (2004). The centrality of culture and community to participant learning at and with the Math Forum. In Barab, S., Kling, R., & Gray, J. H. (Eds.), *Designing for Virtual Communities in the Service of Learning* (pp. 181–209). New York: Cambridge University Press.

Renninger, K.A., Ray, L., Luft, I., & Newton, E. (2006). A comprehension tool for mathematics? The Math Forum@Drexel's Online Mentoring Project. In Barab, S. A., Hay, K. E., & Hickey, D.T. (Eds.), *Making a Difference: Proceedings of the 7th International Conference of the Learning Sciences*. Mahwah, NJ: Lawrence Erlbaum.

Renninger, K. A., Cai, M., Lewis, M. C., Adams, M. M., & Ernst, K. L. (2011). Motivation and learning in an online, unmoderated, mathematics workshop for teachers. *Educational Technology Research and Development* 59(2): 229–247.

Rheingold, H. (2000). *The Virtual Community: Homesteading on the Electronic Frontier*, 2nd revised edition. Cambridge, MA: MIT Press.

Rittle-Johnson, B., & Alibali, M. W. (1999). Conceptual and procedural knowledge of mathematics: Does one lead to the other? *Journal of Educational Psychology* 91(1):175–189.

Rogoff, B. (2005). *The Cultural Nature of Human Development*. Oxford: Oxford University Press.

Rogoff, B., & Lave, J. (Eds.) (1984). *Everyday Cognition: Its Development in Social Context*. Cambridge, MA: Harvard University Press.

Said, E. (1983). *The World, the Text and the Critic*. Cambridge, MA: Harvard University Press.

Sassen, S. (2001). *The Global City: New York, London, Tokyo*, 2nd ed. Princeton, NJ: Princeton University Press.

Schiller, D. (2000). *Digital Capitalism: Networking the Global Market System*. Cambridge, MA: MIT Press.

Schlager, M. S., & Fusco, J. (2004). Teacher professional development, technology, and communities of practice: Are we putting the cart before the horse? In Barab, S., Kling, R., & Gray, J. H. (Eds.), *Designing for Virtual Communities in the Service of Learning* (pp. 120–153). New York: Cambridge University Press.

Sfard, A. (2001). *Learning mathematics as developing a discourse*. In Speiser, R., Maher, C., and Walter, C. (Eds.), *Twenty-First Annual Meeting of the North American Chapter of the International Group for the Psychology of Mathematics Education* (pp. 23–44). Columbus, OH: Clearinghouse for Science, Mathematics, and Environmental Education.

(2008). *Thinking as Communicating: Human Development, the Growth of Discourses, and Mathematizing*. New York: Cambridge University Press.

Sherin, M. G., Jacobs, V. R., & Philipp, R. A. (2011). *Mathematics Teacher Noticing: Seeing through Teachers' Eyes*. New York: Routledge.

Shulman, L. S. (1986). Those who understand: Knowledge growth in teaching. *Educational Researcher* 15(2):4–14.

(1987). Knowledge and teaching: Foundations of the new reform. *Harvard Educational Review* 57(1):1–22.

Shumar, W. (2009). Communities, texts, consciousness: The practice of participation at Math Forum. In Falk, J., & Drayton, B. (Eds.), *Creating and Sustaining Online Professional Learning Communities*. New York: Teacher College Press.

Shumar, W., & Madison, N. (2013). Ethnography in a virtual world. *Ethnography & Education* 8(2):255–272.

Shumar, W., & Renninger, K. A. (2002). Introduction: On conceptualizing community. In Renninger, K. A., & Shumar, W. (Eds.), *Building Virtual Communities*. New York: Cambridge University Press.

Shumar, W., & Robinson, S. (forthcoming). Rethinking the Entrepreneurial University for the 21st Century. In Barnett, R., & Peters, M.A. (Eds.), *The Idea of the University*: Volume 2 – Contemporary Perspectives. New York, NY: Peter Lang.

Shumpeter, J. (1961). *The Theory of Economic Development: An Inquiry into Profits, Capital, Credit, Interest, and the Business Cycle*. Oxford: Oxford University Press.

Silverman, J., & Thompson, P. (2008). Toward a framework for the development of mathematical knowledge for teaching. *Journal of Mathematics Teacher Education* 11:499–511.

Slotta, J. D., & Linn, M. (2009). *WISE Science: Web-Based Inquiry in the Classroom*. New York: Teachers College Press.

Smith, M. A. (2002). Invisible crowds in cyberspace: Mapping the social structure of the Usenet. In Kollock, P., & Smith, M. A. (Eds.), *Communities in Cyberspace*. New York: Routledge.

Soja, E. W. (1996). *Thirdspace: Journeys to Los Angeles and Other Real-and-Imagined Places.* Cambridge, MA: Blackwell.

Spinoza, C., Flores, F., & Dreyfus, H. L. (1997). *Disclosing New Worlds: Entrepreneurship, Democratic Action and the Cultivation of Solidarity.* Cambridge, MA: MIT Press.

Stahl, G. (2006). *Group Cognition: Computer Support for Building Collaborative Knowledge.* Cambridge, MA: MIT Press.

Strange, S. (1986). *Casino Capitalism.* Oxford: Basil Blackwell.

Strauss, C., & Quinn, N. (1997). *A Cognitive Theory of Cultural Meaning.* Cambridge: Cambridge University Press.

Suthers, D. D. (2006). Technology affordances for intersubjective meaning making: A research agenda for CSCL. *International Journal of Computer-Supported Collaborative Learning* 1(3):315–337.

Thomas, J. N., Eisenhardt, S., Fisher, M. H., Schack, E. O., Tassell, J., & Yoder, M. (2015). Professional noticing: Developing responsive mathematics teaching. *Teaching Children Mathematics* 21(5):294–303.

Turkle, S. (1984). *The Second Self: Computers and the Human Spirit.* Cambridge, MA: MIT Press.

(1995). *Life on the Screen: Identity in the Age of the Internet.* New York: Simon & Schuster.

Vygotsky, L. (1986). *Thought and Language.* Cambridge, MA: MIT Press (originally published in 1934).

Weber, S. (1997). The end of the business cycle? *Foreign Affairs* 76(4):65–82.

Wellman, B. (2001). Physical place and CyberPlace: The rise of personalized networking. *International Journal of Urban and Regional Research* 25.

Wellman, B., & Haythornthwaite, C. (Eds.) (2002). *The Internet in Everyday Life.* Oxford: Blackwell.

Wellman, B. Salaff, J., Dimitrova, D., Garton, L., Gulia, M., & Haythornthwaite, C. (1996). Computer networks as social networks: Collaborative work, telework, and virtual community. *Annual Review of Sociology* 22:213–238.

Wenger, E. (1998). *Communities of Practice: Learning, Meaning, and Identity.* Cambridge, UK: Cambridge University Press.

Wertsch, J. V. (1991). *Voices of the Mind: A Sociocultural Approach to Mediated Action.* Cambridge, MA: Harvard University Press.

(2002). *Voices of Collective Remembering.* Cambridge: Cambridge University Press.

(2008). The narrative organization of collective memory. *Ethos* 36(1):120–135.

Wilson, S. M., & Peterson, L. C. (2002). The anthropology of online communities. *Annual Review of Anthropology* 31:449–467.

Wittgenstein, L. (1953). *Philosophical Investigations.* New York: Macmillan.

Wortham, S. (2006). *Learning Identity: The Joint Emergence of Social Identification and Academic Learning.* Cambridge: Cambridge University Press.

Zuboff, S. (1988). *In the Age of the Smart Machine: The Future of Work and Power.* New York: Basic Books.

Index

neoliberalism, 61–2, 66–9, 167
  education and, 70–1
  Math Forum and, 76
  self and, 162–4
Netscan, 35–6
news feed, 68
Noble, David, 71
norms, cultural, 87
Noticing & Wondering (N&W), 11–12,
    54–6, 115, 117–22, 160
  development of, 119
  double reflection and, 15, 118–19, 169
  EnCoMPASS and, 128
  engagement in, 120
  interpretation phase of, 123
  origin of, 118–19
  as reification, 119–20
  rubric of, 128
  scenarios, 121–2
  technology and, 121, 129–30
  theorizing, 122–5
  TPN and, 122–4
NSDL. *See* National Science Digital Library
N&W. *See* Noticing & Wondering

objectification, 76
  Sfard on, 94
OMG. *See* Online Mentoring Guide
OMP. *See* Online Mentoring Project
online communities, 1–2
  of Math Forum, 3–4
online courses, 71–2
online gaming, 5
online learning, 168–70
Online Mentoring Guide (OMG), 104–6
Online Mentoring Project (OMP), 15, 53–4,
    104, 106
OPEC, 62
otherness, 5

Papacharissi, Z., 157–9
participation
  ecologies of, 161
  reification and, 58, 77–8, 80–1, 113,
    127–8
  technology and, 113
Pea, Roy, 50
pedagogic content knowledge, 124
pedagogic thinking, 19–20
pedagogy, Math Monday, 86
Peirce, C. S., 10, 30–1, 80

persistence, 132–4
  Boellstorff on, 147–8
personal webpages, 26–7
Philadelphia, Math Forum in, 45–9
Philadelphia Temple Partnership, 160–2
phone answering machines, 6
physical manipulables, 50–1
physical space, 12
  mediation of, 24–5
politics, of cultural representation, 32
Poster, M., 8–9, 72
postmodernism, 74–5
PoW. *See* Problem of the Week
practices, discourse and, 29–30
pre-service teachers, mentoring, 103–9
Problem of the Week (PoW), 9, 14–15,
    19–20, 27, 30
  archives, 141
  development of, 39–40
  double reflection in, 108
  library, 104–5
  mentoring in, 53–4, 101–2
  monetizing, 44, 75–6
  pupil responses, 108
  resources from, 169–70
  scaling of, 41–2
  scoring, 59, 88–94, 102–3
  technology, 109–10, 121, 143–4
  text in, 140
  wiki, 84–5
problem scenarios, 107
problem solving, 107–8
protean possibilities, 74–5, 163–4
psychology, 29
public sphere, 135–6

Q&A services, Math Forum, 98–9
Quakers, 12–13

Rabinow, P., 21
race, 159–62
RCM. *See* Responsibility Centered
    Management
reflection, 103. *See also* double reflection
  Wenger on, 78
Reflections on Fieldwork in Morocco (Rabinow),
    21
reflexive practice, 116–18
reification, 10–11
  capitalism and, 58
  cyberspace as, 133

CPSIA information can be obtained
at www.ICGtesting.com
Printed in the USA
BVHW071311100619
550590BV00006B/418/P

9 781316 503676